The statistical consultant in action

EDITORS: D. J. HAND AND B. S. EVERITT

The statistical consultant in action

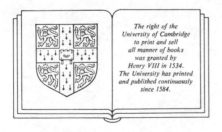

The right of the
University of Cambridge
to print and sell
all manner of books
was granted by
Henry VIII in 1534.
The University has printed
and published continuously
since 1584.

CAMBRIDGE UNIVERSITY PRESS

Cambridge

London   New York   New Rochelle

Melbourne   Sydney

Published by the Press Syndicate of the University of Cambridge
The Pitt Building, Trumpington Street, Cambridge CB2 1RP
32 East 57th Street, New York, NY 10022, USA
10 Stamford Road, Oakleigh, Melbourne 3166, Australia

© Cambridge University Press 1987

First published 1987

Printed in Great Britain by the University Press, Cambridge

British Library cataloguing in publication data
The Statistical consultant in action.
1. Mathematical statistics
I. Hand, D. J.   II. Everitt, Brian
519.5   QA276

Library of Congress cataloguing in publication data
The Statistical consultant in action.
Bibliography
1. Statistical consultants. 2. Statistics.
I. Hand, D. J.   II. Everitt, Brian.
QA276.17.S73   1987   001.4′22   86-17159

ISBN 0 521 30717 1

Contents

Preface

In an article published in 1970 Professor A. Ehrenberg made the following comment:

> 'I feel that the kind of examples of statistical analysis that tend to be considered in professional discussions...are so grossly over-simplified as to make a pretentious mockery of real-life situations and statistical consultancy.'

Examining some of the examples presented in many statistical textbooks, we were forced to agree with Professor Ehrenberg – a comparison of the average weekly beer consumption of male college students living in halls-of-residence with those not living on campus hardly gives the flavour of most real-world consultancy problems!

So in an attempt to present the reality of statistical consulting we were led to assemble the collection of papers that make up this text. Our aim was to show that, in addition to statistical expertise of the type acquired from standard textbooks and in courses held up and down the country, the consultant statistician needs a number of other talents, foremost of which is the ability and willingness to communicate with researchers in other areas. We see this book as a complement to standard texts and ideally we would like to see it adopted as a companion volume on statistics courses, covering those parts of statistical consultancy which are not (and perhaps cannot be) formally taught.

Since statistical problems arise in almost all areas of science, in medicine, in industry, in marketing and in finance, so do problems related to statistical consultancy. Adequate coverage of all these areas could only be achieved in a text containing many hundreds of articles. Nevertheless in the twelve contributions enclosed here we hope a relatively wide spectrum of interests is catered for, and we believe that the problems identified and the general comments made will be applicable in areas not specifically dealt with.

We have attempted to order chapters so that those dealing with similar areas occur close to one another, although this is not always possible simply because several chapters deal with more than one area of application.

The opening chapter presents an overview of the world of the statistical consultant, exploring how different areas of work and different working environments influence the opportunities and problems.

In Chapter 2 Tony Greenfield reminds us that one major difference between British and Japanese industry is that in Japan many millions of people are trained in the use of statistical methods. He presents a case for greater employment of statistical consultants in manufacturing industry and provides an entertaining series of anecdotes about the difficulties which a consultant statistician can encounter.

Professor Barnett, in Chapter 3, presents three case studies, each of which began with a request to 'just help me to fit a straight line to these data'. One is in archaeology and two are in medicine. He uses these examples to show how real problems motivate methodological research, and notes that few subjects can advance nowadays without the proper use of statistical methods.

Crossover designs play a very important role in medical applications of statistics, especially in clinical trials. Clayton and Hills (Chapter 4) present an introduction to the merits and problems of such designs, and then show how theoretical ideals are adapted to practical necessities by examining the analysis of a real crossover design. They also give advice on preparation of the report, noting that different recipients – here a pharmaceutical company and clinical investigator – will expect different things in the report.

In Chapter 5 Cook and Pocock describe some of the ethical issues which arose during a clinical trial, and illustrate the necessity of being able to explain the analyses to the client.

Graham Dunn, in Chapter 6, shows how consultation can be aimed at educating the client, so enabling them to conduct future analyses on their own. He points out the importance of providing an analysis that the client can understand.

Not all consultations are ideal from the statistician's viewpoint. Jeffers (Chapter 7) presents a case in which the consultant statistician had to do his best with data arising from an experiment which was designed and conducted without seeking statistical advice. The case in point is a study of the topical subject of acid rain, and demonstrates how vital it is for the statistician to have some specialist knowledge of the subject matter.

In Chapter 8 Gower and Payne demonstrate how problems presented to the statistical consultant can stimulate the development of both methodology and computer programs.

Lewis (Chapter 9) describes how an unusual outlier problem arose during an investigation of uneven sex ratios in moths. A follow-up on the moths' descendants shows that one cannot afford to be too dogmatic in presenting the results to the clients.

In Chapter 10 Morgan, North and Pack describe collaboration between university and industry, pointing out the special requirements of industrial statistical consultancy, and showing how effective such collaboration can be. They illustrate the benefits which can following from having a consultancy unit attached to a university department of statistics.

Altham, in Chapter 11, describes a problem from manufacturing industry, and demonstrates how simulation can be effective in communicating with the client.

Earlier chapters have described relatively small case studies. Aitkin and Healey (Chapter 12) complement these by describing the management of a study of modelling a large body of socio-economic data.

Finally, for those who wish to pursue further any of the non-statistical aspects of consulting, we present a bibliography on statistical consultancy.

We hope that the material in this book will show the budding statistician that statistical consultancy work provides a very broad spectrum of interesting and intellectually stimulating problems. The working environments are equally varied, ranging from university mathematics departments, via specialist advisory units, to freelance industrial and commercial work. Few other professions provide one with the opportunity to contribute significantly to progress in such a wide diversity of disciplines.

Lastly it is a pleasure to express our sincere thanks to all our contributors for the speed with which they responded to our invitation to submit articles and for their forebearance when, where necessary, they were reminded of 'deadlines' (realistic or not!).

D. J. Hand
B. S. Everitt

Contributors

Professor M. Aitkin, Centre for Applied Statistics, University of Lancaster, Bailrigg, Lancaster, LA1 4YL

Dr P. M. E. Altham, Statistical Laboratory, University of Cambridge, 16 Mill Lane, Cambridge, CB2 1SB

Professor V. Barnett, Department of Probability and Statistics, The University, Sheffield, S3 7RH

Dr D. Clayton, Department of Community Health, Clinical Sciences Building, Leicester Royal Infirmary, PO Box 65, Leicester, LE2 7LX

Dr D. G. Cook, Department of Clinical Epidemiology and General Practice, Royal Free Hospital School of Medicine, London NW3 2PF

Dr G. Dunn, Biometrics Unit, Institute of Psychiatry, De Crespigny Park, London, SE5 8AF

Mr B. S. Everitt, Biometrics Unit, Institute of Psychiatry, De Crespigny Park, London, SE5 8AF

Mr J. C. Gower, Statistics Department, Rothamsted Experimental Station, Harpenden, Herts., AL5 2JQ

Dr A. A. Greenfield, The Heights, Bradway, Whitwell, Herts, SG4 8BE

Dr D. J. Hand, Biometrics Unit, Institute of Psychiatry, De Crespigny Park, London, SE5 8AF

Mr R. Healey, Savant, Carnforth, Lancashire

Dr M. Hills, Department of Epidemiology and Statistics, London School of Hygiene and Tropical Medicine, Keppel Street (Gower Street), London WC1E 7HT

Dr J. N. R. Jeffers, Institute of Terrestrial Ecology, Merlewood Research Station, Grange-over-Sands, Cumbria, LA11 6JU

Professor T. Lewis, Department of Statistics, The Open University, Walton Hall, MK7 6AA

Dr B. J. T. Morgan, Mathematical Institute, Cornwallis Building, The University, Canterbury, Kent, CT2 7NF

Dr P. M. North, Applied Statistics Research Unit, Mathematical Institute, Cornwallis Building, The University, Canterbury, Kent, CT2 7NF

Mr S. E. Pack, Mathematical Institute, Cornwallis Building,
The University, Canterbury, Kent, CT2 7NF
Mr R. W. Payne, Statistics Department, Rothamsted Experimental
Station, Harpenden, Herts., AL5 2JQ
Dr S. J. Pocock, Department of Clinical Epidemiology and General
Practice, Royal Free Hospital School of Medicine, London, NW3 2PF

1

Statistical consultancy

D. J. HAND AND B. S. EVERITT

1 Introduction

Since this is a book about statistical consultancy, a good place to start would be by considering the ideal consultation. Hyams (1971) has given us a description of this:

> To qualify a consultation as ideal is to deny its empirical meaning. The 'Ideal Consultation' is *not* a consultation. It is a working-together, a voluntary meeting of minds and union of energies whose prime aim is to seek a 'truth'. In such meetings both parties are familiar with each other's basic language. The biologist has had a few courses in basic statistics and thus recognises statistics as a unique and valuable discipline. The statistician has also done his homework and has familiarised himself with the names and the relationships of the fauna in the experimenter's jungle. Since knowledge and understanding breed sympathy and respect, the researcher esteems the statistician as an expert representative of this most important science. His appreciation for the statistician's unique contributions grows by leaps and bounds with the experience of his individual talents. Needless to say, the feeling is mutual. Meetings are stimulating; they are productive in thought and in product. The work forms a gestalt (the whole is greater than the sum of its parts). The research poses challenging statistical problems that are fun to work at: the sort of thing that keeps one busy at a scratch pad during supper while the wife silently suffers (or throws a fit). In unhurried time the deliberations proceed to a design, an experiment, and an analysis that confirms everyone's best hopes. The (multiple) reports are easy to write. Sometimes the biologist's name is first, sometimes the methodologist's; it hardly matters.

These manuscripts are received enthusiastically by journal editors and their 'expert' reviewers don't give the team a hard time. First experiments lead naturally to others and the information generated finds a significant practical application. Ultimately it saves human lives or curtails misery. Finally, but justly, the co-workers are awarded the Nobel Prize in Medicine and quite naturally donate their stipends to schools encouraging interdisciplinary approaches to problem solving.

In practice there seem to be some slight (!) differences between this ideal and the reality, so a natural question to ask is where are we going wrong, or more realistically, what problems do consultant statisticians have to contend with which disrupt such an ideal scenario?

2   The problems of statistical consultancy

The primary issue affecting realisation of the above ideal is, of course, the relationship between the client and the consultant. This will naturally depend on personalities, preconceptions of the role of the other party, and the nature of the relationship between the individuals concerned. We shall have more to say about these things below, but the sort of factors involved are: how much statistics the client knows; whether he is prepared to accept advice or is confident he knows the answers and is just seeking confirmation; whether the consultant is a freelance statistician or a junior member of a university department of which the client is the head; whether it is to be a genuine collaboration or whether the statistician will merely be acknowledged in some subsequent paper (and does he want this anyway, if the client has ignored his advice?). Perhaps most important of all is the question of whether the statistician is perceived as a scientist in his own right. This sets the tone of the relationship.

A number of authors have examined the relationship by condensing their experience of clients into a classification. Sprent (1970), for example, produces the following seven classes:

1. *The timid apologist* who has little statistical knowledge and expresses reluctance at wasting our time. It takes something akin to a doctor's bedside manner to put such people at their ease and overcome their reluctance to seek help.

2. *Significant difference and least significant difference (l.s.d.) experts.* These are encouraged by misguided editors who think all numerical results can be made respectable by quoting significant differences or significance levels – often denoted by * or ** or ***, a symbolism more appropriate to a hotel guide-book than a serious scientific paper. The number of editors accepting l.s.d.'s

or asterisks as the ultimate in statistical sophistication is happily declining. In my experience it is a characteristic of inveterate users of significance levels that they do not know what they mean. Once the meaning is explained to them more sophisticated ideas appeal to them. The number of l.s.d. experts might decrease if we pressed for less emphasis on significance in elementary service courses.

3. *The one-technique amateur statisticians.* These are proficient at just one technique and show great ingenuity at applying it even when it is not relevant. It is worth some effort to expand their statistical horizons.

4. *The believers in sacred texts or computers,* have been mentioned in the previous section.

5. *Experimenters with addled statistical ideas.* These are the people who assure you that, for example "the object of random-isation is to disperse treatments as widely as possible over the experimental area," or that they "never use randomised blocks, but always do factorial experiments instead". It requires some diplomacy to replace such misconceptions by useful knowledge.

6. *The expert data handler who is not a statistician.* Such a person may be described as a born data handler who relies upon his instincts when handling data. It is usually easy for the statistician to collaborate with him, but one feels that he spends perhaps too much time re-inventing known techniques if he proceeds without some statistical help.

7. *The statistically informed experimenter* is always a pleasure to work with. Not only does he understand our jargon, but he usually sees to it that we understand that of his subject so far as is necessary.

Hyams' (1971) classification is also worth reading, yielding the categories: probabilist, numbers collector, sporadic leech, amateur statistician and long distance runner.

(If, at this point, the reader should feel we are being unfair on the client in what is, after all, a two-party relationship, we hasten to reassure you that the balance will be redressed below.)

We mentioned, above, the lack of statistical knowledge on the part of the client. It is precisely because he lacks this knowledge, and is aware of it, that he is calling upon the professional services of the statistician. A more subtle source of potential pitfalls, however, lies in lack of expertise the other way round. How much does the statistician know about the client's discipline? Inadequate knowledge can not only lead to difficulties for the statistician in formulating the researcher's questions in a way he can answer, but it can lead to fundamental misconceptions of the researcher's aims. Moses and Louis (1984) give an elegant little example of this, illustrating the importance of understanding what the measurements mean and how they were obtained: the statistician is presented with two measurements of phase angle, 10° and 350°, and works with their average of 180°. Because of these sorts of difficulties Cox (1968) has suggested that there should be texts describing other scientific disciplines specifically aimed at statisticians, just as there are statistics texts aimed at other scientists. Indeed, recognising this symmetry of the client/consultant relationship, Sprent (1970) concludes his taxonomy of clients by saying: 'Whether or not an experimenter fits neatly into one of the above categories, a collaboration will never be helped by our sneering at his statistical ignorance. I shudder to think how often I have appeared naive when talking to experts in another field about their speciality.'

Several authors take the symmetry further, and suggest that one of the roles of the consultant statistician should be as teacher, educating the researcher in statistical methodology. From this it also follows that what is an 'optimal' solution in practice may differ from what is 'optimal' in theory. A complex multivariate repeated measures analysis of variance on logged data may be perfect from the theoretical statistician's viewpoint, but if the client has no hope of understanding it then it is worse than useless – worse because of the reaction it will provoke and the role misconceptions it will create.

A problem which seems ubiquitous in modern life is lack of time. In statistical consultancy this can manifest itself in two ways. The first is the client's lack of time: he needs an answer by tomorrow at the latest. This might be because he is under pressure from his boss, because his business will collapse if an answer is not available, or because he is presenting the results at a conference the day after. It might simply be that he knows that computers are immensely fast and so does not see any problem in producing a result within a few minutes or hours at most. Of course, apart from the basic misconceptions in this notion, it also reflects poorly on the client's understanding of the other pressures on the statistician. This is the second kind of time pressure: the statistician's lack of time. For statisticians who function in a service capacity this can be a very serious problem. Often there are ten other clients queuing outside the door while one is grappling with the intricacies of some complex data set. The temptation, of course, is simply to adopt the most straightforward procedure as a solution, and then have to live with the feeling that one could have done so much better if only one had had the time to consider the problem properly.

Often the complex data set referred to above was collected without first seeking the advice of a statistician. Not infrequently years of effort have gone into collecting the data. This complaint, that the statistician was called in too late, is a common one. At its worst it can lead to the waste of vast sums of money or the abandonment of a PhD. Helen Kraemer (quoted in Moses and Louis, 1984) says:

> If consultation is at the post hoc stage, it may be that the objectives cannot be accomplished (sampling bias, poor design, etc.). It is the statistician's responsibility to state this frankly. We cannot do magic, and we can't participate in cover-ups. It is as well that researchers know our limitations in advance. This is a particular problem when the first consultation takes place after a research paper is rejected for publication because of poor methodology. Not much one can do!

(It is interesting, however, that Daniel (1969) says: 'Some statisticians would say that the only favourable time to enter a research project is at its beginning. My own experience does not confirm this. I have entered projects at all stages of their development. I do not see any connection between my stage of entry and my success or failure.')

Another pressure influencing the consultant/client relationship is the less obvious one of ethical issues. These are, of course, well known in the medical field and perhaps also in social applications of statistics. But they occur elsewhere as well. An example would be in a university environment, where moral problems can arise with the students seeking help: just how much statistical advice should one provide? Whose PhD is it anyway?

3   A taxonomy of statisticians

We have presented, above, a typology of the client. In fairness we must also present one such of the consultant. Hyams (1971) gives us the following:

> 1. *The Model Builder* fits any and every data problem set to a model he is presently interested in or knows something about. It matters not whether he investigates the questions that are being asked by the client or those that are biologically important. For that matter, this type isn't really interested in hearing the client's story. He had posed his own a priori questions before the client knew him. The Model Builder is like the drunkard looking for his lost key under the street lamp although he dropped it in the dark alley. He justifies his search by pointing out that there is light in the place he is looking.
> 2. *The Hunter* is the statistician counterpart of the Numbers Collector who directs you to 'mine the mountain'. The Hunter

will subject every data set to an exhaustive and extensive computer analysis. For a relatively simple problem with scanty data he will ultimately present the investigator with 14 vertical inches of print-out, containing 17 significant results. These numbers do not bear a relationship to anything on the face of this earth except themselves. While the client may initially accept these authoritative materials with reverence, it will not take him long to figure out that he has a bag of wind.

3. *The Gong* is a consultant who starts every conference by drawing a bell-shaped curve.

4. *The Traditionalist* is convinced that nothing really important has happened in statistics since R. A. Fisher and consequently limits himself to a restrictive working vocabulary. He views the computer as the devil's work.

5. *The Randomophiliac* firmly believes that it doesn't matter what else you do, as long as you've randomised well. He is like the mother who catches her 14 year old daughter in a sexually compromising situation and admonishes her by saying "as long as you don't smoke, honey".

6. *The Quantophreniac's* position is: It doesn't matter if you observe what you want to so long as you get a hard measurement.

7. *The More Data Yeller* (he needs no further description).

8. *The Nit Picker* will always focus his attention on the inconsequential but debatable. He will enlarge minor issues out of reasonable perspective and quickly reduce a real and tremendous contribution to a potentially horrendous error in reality testing. (My manuscripts are usually reviewed by this type.)

To these we might add the *problem stealer* (who decides that every problem would make a perfect project for his students, to be begun next summer; from this G. J. Goodhardt (1970) derives his 'rule 1 of the business – never consult an academic in October') and the *allied problem solver* (who, as Goodhardt (1970) says: 'finds great interest, not in my problem, but in some other problem that mine suggested to him. This may take the form of a wider generalisation of the conditions which unfortunately does not happen to include the special case I started with, or a detailed description of the intricacies of estimation in small samples when I have a sample of size 5000.').

One hopes that this is just a list of inadequate types, the good and competent ones having been omitted from the list.

These caricatures at least make it clear what we should strive to avoid. Presented with them one might justifiably ask how we should go about training effective consultants. This question is dealt with in a number of places (for example, Committee on Training of Statisticians for Industry, 1980; Boen, 1972; Griffiths and Evans, 1976; Tarter and Berger, 1972; Watts, 1970; Zelen, 1969; and Zahn and Isenberg, 1983). This book is, we hope, a further answer, complementing the advice on training given in the above by exposing the reader to a taste of the wide range of problems that will be encountered in the life of a statistical consultant.

4   Statistical domains

It has been suggested that after one's formal training in statistics (to BSc, MSc or PhD level) it then takes a further three years functioning within a particular application environment before one attains sufficient competence to act as an independent consultant. The reason for this will be partly the need to acquire the personnel skills mentioned above, partly the need to adjust to the problems of real data (missing values, outliers, multiple sources, etc.), and partly the fact that different areas of application place different degrees of emphasis on different techniques. The extent to which this is true is illustrated by the fact that, even within statistics itself, the technical term 'theory of reliability' has two quite distinct meanings. One refers to the reliability of (for example) complex machines, and the other to the consistency with which measuring scales yield identical results (in the behavioural sciences).

Application domain is just one type of categorisation which can be used to describe statistical consultancy work. A second is the working environment. For example, the statistician might be an academic who spends a small part of his time (voluntarily) advising people; or he might work from a service unit, with his primary function being to advise; or he might be a freelance consultant, who eats or goes hungry according to the success of his consultancy work. These three types have very different roles and requirements.

The academic can afford to look merely at interesting (to him) problems, can afford to be sidetracked to more interesting ones (if a client does not return for more advice it does not personally damage the statistician), and he may not be part of a team.

The statistician within a service unit is obliged to answer, or attempt to answer, the questions of anyone who knocks at his door. There is a danger that the role of the statistician, as a scientist in his own right, will not be properly perceived. Armitage (1970) says that as much as possible of the service function should be handed over to the client himself. (In fact Armitage prefers the term 'advisory work' unless there is a commercial agreement, a point with which we agree.)

For statisticians who fall into either of the above categories it is essential for precise roles to be mapped out beforehand. Is it to be a collaboration, with both names appearing on any subsequent publications?

In contrast, for statisticians in the third class – that of the freelance consultant – the roles are already well delineated. The financial motivation sees to that.

5   Computers and statistical consultancy

Computers, of course, have revolutionised statistics. How many of the case studies in this book would have been feasible without them? But the real potential of computers is only just beginning to be realised.

The initial impact of computers was to speed up, to minutes, techniques which previously would have taken days or weeks to apply by hand. This has had the consequence of much more widespread application of these methods and of pushing the applications further to bigger and more complex problems. Much more interesting than this, however, has been the development of new techniques for which the computer is absolutely essential and which, without computers, just would not exist. Examples of such children of the computer age are log-linear modelling, kernel density estimation techniques, and bootstrap methods. There is no evidence that progress in this direction has stopped. Developments are continuing, and the advent of even more potent computers makes the prospects truly exciting.

Apart from the development of new techniques, developments in computers seem likely to revolutionise statistical consultancy work from a different direction, and one which could not have been predicted before the computer age. This is that of interactive statistical graphics. The approach to data analysis in laboratories with access to fast and high-powered interactive graphics facilities is diverging from the more traditional approach.

Returning from the frontiers to the more mundane, we find widespread access to powerful statistical packages such as SPSS, SAS, BMDP, etc. Such packages are easy to use. This means that they can be used by those who are relatively untutored in statistics. They can equally easily be misused. Hooke (1980) says: 'Use [of statistics] has been replaced by overuse and misuse. Regression is being used in foolish ways in the neighbourhood of almost every computer installation.' And the rate of errors in published analyses suggests that the problem is serious (see White, 1979; Gore, Jones, and Rytter, 1977; and Altman, 1982).

We can hardly impose a moratorium on the use of such packages by those who are not professionally qualified, so this has motivated the development of statistical expert systems, systems which statistically inexperienced researchers can use to analyse data and which will protect them from error. (See, for example, Gale (1985), Hand (1984, 1985), Pregibon and Gale (1984). One of the earliest references to the possibility of building this kind of statistical expertise into packages is Finney (1970).) Whether such systems will serve statistics by preventing some of the criticism which is currently misdirected at it, rather than at those who misuse it, remains to be seen. In any case, it will clearly be a long time before such systems can handle the kinds of problems to which this book is addressed: that is, the problems which lie at the interface between statistical expertise and expertise in the discipline of the client. Resolution of such problems requires not only statistical knowledge, but also wider knowledge of the world and the way it behaves.

6   Conclusion

The problems facing statisticians serving as consultants are varied: varied not only in the origin of the data and the research questions presented, but also in the kind of personal skills they will require the statistician to possess in order to resolve the questions successfully. Communicating with statistically and mathematically naive research workers can be an exacting and, on occasions, a frustrating task, and patience and tolerance are likely to be needed in good measure. Nevertheless, working as part of a team to solve practical problems can be very exciting, and the intellectual rewards great. The statistician as an expert on the formulation and manipulation of mathematical models and on research methodology is in an ideal position to act as a catalyst in drawing together members of a research team. In this central role the statistician is far more than merely a second class mathematician.

To become successful consultants, students clearly need to acquire a grasp of the practical problems they will encounter, in addition to the theoretical expertise imparted by their courses. It is hoped that the diverse range of real problems described in this collection will go some way towards filling that need.

7   Further reading

At the end of this book we present a bibliography of work on the practical aspects of statistical consultancy. General works which the reader might find interesting are those by Sprent (1970), Hyams (1971), Feinstein (1970), and the book by Boen and Zahn (1982).

References

Altman, D. (1982) Statistics and ethics in medical research: VIII. Improving the quality of statistics in medical journals. *British Medical Journal*, **282**, 44–7.

Armitage, P. (1970). Discussion following Sprent (1970).

Boen, J. R. (1972) The teaching of personal interaction in statistical consulting. *The American Statistician*, **25**, 30–1.

Boen, J. R. and Zahn, D. A. (1982) *The Human Side of Statistical Consulting*, Belmont, California: Wadsworth.

Committee on Training of Statisticians for Industry (1980) Preparing Statisticians for careers in industry (with discussion). *The American Statistician*, **34**, 65–75.

Cox, C. P. (1968) Some observations on the teaching of statistical consultancy, *Biometrics*, **24**, 789–802.

Daniel, C. (1969) Some general remarks on consultancy in statistics. *Technometrics*, **11**, 241–6.

Feinstein, A. R. (1970) Clinical biostatistics VI: statistical malpractice – and the responsibility of the consultant. *Clinical Pharmacology and Therapeutics*, **11**, 898–914.

Finney, D. J. (1970) Discussion following Sprent (1970).

Gale, W. A. (1985) *Artificial Intelligence and Statistics*. Reading, Massachusetts: Addison-Wesley.

Goodhardt, G. J. (1970) Discussion following Sprent (1970).

Gore, S. M., Jones, I. G. and Rytter, E. C. (1977) Misuse of statistical methods: critical assessment of articles in BMJ from January to March 1976. *British Journal*, **1**, 85–7.

Griffiths, J. D. and Evans, B. E. (1976) Practical training periods for statisticians. *The Statistician*, **25**, 125–8.

Hand, D. J. (1984) Statistical expert systems: design. *The Statistician*, **33**, 351–69.

Hand, D. J. (1985) Statistical expert systems: necessary attributes. *Journal of Applied Statistics*, **12**, 19–27.

Hooke, R. (1980) Getting people to use statistics properly. *The American Statistician*, **34**, 39–42.

Hyams, L. (1971) The practical psychology of biostatistical consultation, *Biometrics*, **27**, 201–12.

Moses, L. and Louis, T. A. (1984) Statistical consulting in clinical research: the two way street. *Statistics in Medicine*, **3**, 1–5.

Pregibon, D. and Gale, W. A. (1984) REX: an expert system for regression analysis. *COMPSTAT-84*, Czechoslovakia.

Sprent, P. (1970) Some problems of statistical consultancy (with discussion). *Journal of the Royal Statistical Society*, Series A, **133**, 139–65.

Tarter, M. and Berger, B. (1972) On the training and practice of computer science and statistical consultants. *Proceedings of the Computer Science and Statistics Sixth Annual Symposium on the Interface*, American Statistical Association, pp. 16–23.

Watts, D. G. (1970) A program for training statistical consultants. *Technometrics*, **4**, 737–40.

White, S. J. (1979) Statistical errors in papers in 'The British Journal of Psychiatry'. *British Journal of Psychiatry*, **135**, 336–42.

Zahn, D. A. and Isenberg, D. J. (1983) Nonstatistical aspects of statistical consulting. *The American Statistician*, **37**, 297–302.

Zelen, M. (1969) The education of biometricians, *The American Statistician*, **23**, 14–15.

2

Consultants' cameos: a chapter of encounters

(*The aetiology of a statistician's paranoia!*)

TONY GREENFIELD

What is life really like for the consultant faced with the clamour of consulters, engineers, scientists, business managers, doctors, each of whom believes that his is the only problem in the world worthy of your attention? Can you always muster the humour, the assiduity, the indominatability, and the extra ten hours a day that you need to cope with it all? I admit it's difficult and, just as you all have your tales to expose these untutored needs of the job, here are some of mine, linked with the odd word or two of advice. Some of them originate from my time in industry, others from teaching hospitals, but the characters and attitudes are ubiquitous.

It was a lovely day. Clouds bustled billowingly white in the warm summer breeze under the bright blue sky. Too good to be indoors sweating over a hot computer, I thought, as I locked my car and strode towards the office building.

'Ah, just the man,' came an Ulster evangelical bellow. It was a consultant. No, not one of us, but one of them. Genuflecting slightly, I tried to excuse myself, but too late.

'I just want to ask you a quick question,' he said, and spread a file of papers over the nearest car bonnet. The wind grabbed a handful and distributed them among a dozen wheels. The boy scout in me raced around retrievingly and then I realised he had me.

'Perhaps, it's too breezy here,' he said, 'Let's go to your office.'

'I'm sorry, but I already have a client waiting, and there's a heap of analysis that I've promised for others by tomorrow.'

'Don't worry. This will only take a moment.' He led the way to my office, briefly acknowledging the waiting registrar with: 'I'll be out in a minute.'

At last words reached my lips. 'I'm sorry, but no statistical question can be put in just a minute let alone answered. Can you come back next Monday?'

'Oh no, it's much more urgent than that. I have to write the paper at the weekend. It's the only time I've got. But I see you're busy so I'll just leave the file and will call back later.' And without fixing a date he left.

Two weeks passed and I was sitting with another consulter. The big man entered and interrupted: 'Have you looked through that file?'

'Sorry, but I really have been too busy with existing commitments.'

'Yes, yes, but this is quite urgent now. Do have a think about what analysis would be best and I'll leave the data later.'

That evening he interrupted another session. 'There's the data,' depositing a heap of forms. 'I'll look in tomorrow.'

'Just a minute,' I called after him. 'How do you think I could manage if all my clients demanded such an immediate response?'

'Ah, but this isn't just for anyone,' he said. 'It's for me.'

Next week he passed me in the corridor. 'I don't suppose you've had a chance to look at my analysis yet? No? Well I really must make an appointment with your secretary to discuss it with you.'

The file and data forms may still be sitting there yet.

At jazz concerts you may applaud at any moment. Any tricky drum roll or sax trill will do but it is absolutely *de rigueur* if you recognise a long lost favourite melody. The players don't mind. They even like it. The clapping urges them into further fanciful flights. But at an orchestral concert you must keep mum. Not a murmur must leave your lips, nor a plaudit pass your palms until after the final beat of the baton. A delicate cough, no more, may be allowed between movements. But it was just then that my shoulder was tapped. Not the same *the* consultant, but another just the same. 'Lucky I spotted you,' he whispered. 'Can I have a word with you in the interval?'

There are many more just like them.

My office door was closed and clearly marked 'engaged'. I sat behind it, determined not to be disturbed until I'd mastered a promising technique offered in the latest journal by a leading theoretician with a sadly obscure style. (Please note: we applied statisticians are truly grateful for the continuing advances you offer us but wish you would make them easier to understand quickly.) The door opened and the head of yet another said: 'I can see you're alone so I'll ignore the sign.'

The lesson is: employ a secretary who is not only charming as a receptionist but also tough and understanding and have her sit the other side of your office door, not in a room down the corridor. And learn to say 'no'. Not to her of course but to them.

The idea that a statistician is just an artisan who does as he's told and doesn't need to think, came shortly after a new boss was appointed above me in the industrial laboratories. This coincided with the introduction of a new staff grading system which was clearly designed to upset as many people as possible. The new manager quickly made his mark. He was better placed than the old hands to do this because he was able to grade staff according to his snap perceptions of their jobs without being influenced by past personal familiarities. 'Your section provides a technical service,' he said. 'You are not scientists, but technicians. Naturally, as section head, you will be graded as a senior technician.'

Appeal to professional qualifications, job advertisements, published salary surveys and authorship lists in scientific journals, quickly overcame this view and we became friends and, almost, equals. But the incident does introduce an attitude that exists among some of our clients: that we are simply technicians equipped with a bag of tools from which we must draw the right one to tackle the job in hand. Very often the client will even tell us which tool to use.

A man from an engineering research laboratory had read that the F-distribution was that of the ratio of two chi-square variates, so he brought me some chi-square estimates from a set of contingency tables asking me to refer the ratios to my table of the F-distribution. When I demurred, saying that I should prefer first to understand the main research problem before considering how the data should be treated and the estimates tested, he challenged me with: 'You are the statistician, aren't you? [Note the 'the', as if I were 'the' plumber.] Well let's get on with it. I just want a simple test doing and then I won't waste any more of your time.' I am happy to report that he left unsatisfied, but sad that his opinion of me was 'unhelpful'.

Another amateur statistician who knew just what he wanted was a general practitioner who was comparing the incidence and treatment of myocardial infarction in his country town with those of a neighbouring town.

He arranged a meeting with me and asked if two other members of my department could be present. At the agreed hour he trooped in with his entourage and suddenly there were eight of us sitting round the table. He produced his pile of data forms and a list of instructions: the tables to be printed, the tests to be done, and the timetable for the work. Tentative questions about the relevance of his plan surprised him for their effrontery. After all, it was his research and any suggestion for treating the data in a different way implied criticism of his own professionalism: we were there to compute.

The lessons from both these examples are: avoid, if you can, getting involved in any study so late in the day that you cannot influence its definitions and design. Preliminary talks about a study should be between you and the principal researcher only and not in a working group or committee.

One of the advantages of an early meeting with just one person is that it doesn't matter if you know nothing at all about the subject of the study. Indeed it's often an advantage that will enable you to learn much more than if you appeared to be knowledgeable. It will also help to clarify your consulter's own mind. Naivete is strength: admit your ignorance. Insist on a clear description of what it's all about from the most elementary level possible. Simply keep asking questions with a bland look upon your face. Even if you do understand much of the subject in advance, keep it to yourself. Pretend an ignorance. Then occasionally allow some of your knowledge to slip out, but do it as if you had just begun to see the picture, as a tribute to the consulter and his ability to explain. Generally stick to simple questions, like 'What do you hope to achieve?' and 'How will you know if you've done it?' But also show your eagerness to learn some elementary chemistry, physics, biology, or psychology from the consulter. Many a person has left my office after an hour or more of such a consultation, thanking me lavishly for all my help when in fact I have contributed nothing except questions based on real or feigned ignorance. Yet there is a real contribution here: it is forcing the consulter to think clearly in a context that will make sense to those of us who are outside his sphere of specialism.

Letting slip a little knowledge? Sometimes I can't resist the temptation to puncture a client's pressurised ego. The trick is to find his weakness. In the following example he pretended the weakness was mine so as to conceal his own.

The rheumatologist responded well to my air of innocent ignorance. He confessed to being a leading expert and implied that his leadership was global. His gratuitous tutorial on the structure of joints and their pathological processes was increasingly patronising until he said: 'This is where free radicals come into play, but I won't waste time explaining that to you because you couldn't possibly understand the theoretical chemistry.'

'Oh don't worry about that. I'm familiar with the physics of free radicals, especially the superoxides.'

'You are? But none of the biochemists in the hospital know about it!'

'Ah, but I'm a mathematician.'

Thank you, Stephen Potter.

There may even be occasions when, with some courage, it is worth declaring your ignorance publicly. After all, the purpose of any meeting should be to learn from each other and not to show how clever we are. At a meeting to share knowledge about breast cancer between scientists of mixed disciplines, a physician began to lecture us about the technique of oestrogen receptor blockade. Within a few minutes it became apparent that he expected us all to be as familiar as he was with histology, cytology, endocrinology and the biochemistry of RNA mediated protein synthesis. I couldn't take it. So I stood up and, cupping my hands to my mouth, bellowed 'stop' like the man who used to bring London's traffic to a sudden quiet standstill in the old wireless program 'In Town Tonight'.

'I really am sorry to interrupt you,' I told him, 'but I honestly do not understand what you are talking about and I suspect there are others in the audience who don't either.' Heads were nodding in agreement. 'Could you please start again?'

He was a very nice chap and took it very well. He started again with a crystal clear exposition. An hour later not only the statisticians but some of our more medical colleagues agreed they now understood the subject better. I fancy that our lecturer may have done too.

Similarly, ignorance of statistical methods by the consulter is acceptable if only he will admit it as honestly as you admit ignorance of his subject. But it can pose problems. There were three arms in the drug trial: a placebo and the new drug administered with two dosages. The response was measured just before administration and at several subsequent times. Although multiple *t*-testing seems, from what is published in the medical journals, to be standard practice, a comparative time series analysis was indicated and was done. But the client didn't understand it and pleaded with what you must all have heard many times: 'All I want is a *P*-value!' But at least he knew he didn't understand and was willing to be guided through an explanation.

Hyams (1971) told us: 'The consultant should not offer solutions that are beyond the comprehension of the experimenter or his ability to describe them.'

On the other hand, how do we deal with the man who is so ignorant that he is totally confident of his false knowledge? Unless you can refer him to another consultant who will support you, perhaps the only way is to let him sink in his own mire. But then, can you live with the knowledge of the consequences?

An engineer asked me to derive a function relating the tilt of a ladle to the volume of liquid steel remaining in it. The ladle was a truncated cone so, if it were tilted to pour out some steel the surface of the remaining steel would be a conic section. If the edge opposite the pouring edge was touching the ladle wall, the surface must form an ellipse. 'Nonsense,' he said, 'the edges have different curvatures, so the surface must be egg-shaped.' I tried to assure him that a far greater mathematician than me, a chap called Euclid who had lived in Alexandria two millennia before, had settled the matter for all time. But he wouldn't have it and went away. His confidence in me was destroyed and now, somewhere, there is a foundry whose casting yield never quite balances.

I was reminded of Good's aphorism: 'Half-baked ideas of people are better than ideas of half-baked people.'

Others must be admired for their persistence.

Our statistical assistant came in. 'Doctor X wants me to analyse his data to find an association between blood alcohol and sugar levels in a certain class of patient.'

'Have you told him he should see one of the lecturers first?'

'Yes, but they've all told him to go away.'

'Why's that?'

'He has only five observations,'

I couldn't speak.

'Will you see him please,' said John. 'I don't know what to do.'

Doctor X appeared and confessed his story. He'd first approached a member of the department more than six months before. He had had five observations then and had been advised to write out his objectives as a preliminary to designing an experiment and collecting more data. Instead, he had written to the organisers of an international conference offering to present the results of his research. He had then, at intervals of several weeks, visited other statisticians and been given much the same advice.

But he still had only five observations. The international conference was next Thursday and he had 15 minutes allotted to address several thousand specialists.

'You must withdraw immediately,' I told him. 'They'll howl you off the platform.'

'All I want is a simple answer,' he said. 'Why can't you statisticians be more helpful?'

His cussed nature allowed him no retreat, so I *had* to help him. My advice was to show a plot of the five points, to declare that there seemed to be an association but there was nothing at all certain, and to argue that this might constitute a hypothesis on which further research might be based. Simple enough?

Two weeks later I met him in the park. 'You've survived then?'

'It was a triumph,' he said. 'They were all very interested and want me to go back next time with more data.'

Then there are those with too much data.

The phone rang. 'Are you *the* statistician?'

'Well, just one of a few about the place.'

'I've just moved from Birmingham and was told you could help to move my data file.'

'Perhaps I could, but somebody in the computer centre would be better. I'll give you a name and phone number.'

'Well, when that's been done, may I talk to you about the analysis?'

'What do you have in mind?'

'Nothing in particular. I thought we should just look through the data to see if we can spot any relationships.'

'Oh yes! And how much data is there?'

'About 100 variables on each of about 10 000 cases.'

I find it hard to sympathise with, let alone willingly help, a man who has submitted 10 000 patients to detailed examination and questioning and who has, regardless of cost to the nation, printed 10 000 forms, consumed countless clerical hours for coding and data entry, and grabbed so much precious computer time and storage space. Nor could I ignore the damage it might do to my own reputation if it became known that I were associated with such a pointless venture.

And don't believe you can always control the publicity. In another study there were only 99 cases but with 72 variables. Submitting to pressure, I gave the client the correlation matrix he'd begged, but with the caveat that this was only to let him get a feel for the data, that no inferences were to be drawn until hypotheses had been clearly stated, used for the design of an experiment upon which further data would be collected, and estimates and tests properly done. Two years went by and I'd forgotten all about it until a paper arrived on my desk, with his compliments, and me as a co-author. The paper was based on those few correlations which had happened to be high. He really believed he'd been nice to me.

The lesson from all these examples is: Do not embark on any project until the objectives, constraints, costs, and publication proposals are properly documented in a letter, memorandum, or standard job form.

If you do use a standard form, make sure it's watertight. I was asked to design a form for the local medical research ethical committee, to be used by applicants for approval for their projects. Seizing the opportunity to promote the indispensability of statistical consultancy, I included the question: 'Have you had statistical and/or computing advice? If so, from whom?' When the forms had been in use for some months I was reading some applications and came across one that, at best, was dubious. Yet the applicant had named a member of my department, implying that he had approved it from the statistical viewpoint.

'Has this man really consulted you?' I asked the statistician.

'Yes he has. And I advised him not to do it!'

The form has been changed. It now includes the statement: 'If you name a local statistical adviser attach a letter from him.'

My involvement with the medical research ethical committee exposed me to the still persisting belief of some people that it is not the role of the committee to question the scientific basis of a study including its experimental design. You may be familiar with the sentiment: 'This is supervised by Fred. Fred's a very nice fellow and wouldn't hurt a fly so the study will certainly be done ethically.'

It is a role of the consultant statistician to sit on such committees. In other sciences and in industry there are similar committees but their main criterion is the economic management of resources rather than human ethics. But, in either case, the vetting of their people's research can lead to difficult confrontations. A post graduate student attached to an industrial laboratory proposed a study to determine the relationships between various compositional and process variables and the mechanical properties of a certain class of material. His intended tests would take just under three years which, coincidentally, was the time covered by his research grant. When I showed that experimental design would enable better information to be obtained in only three months, the management were pleased but the student was very upset for a while, believing I had sabotaged his career. Fortunately he soon realised that there was more he could do with his time.

But a dispute with a Very Important Person can lead to permanent personal severance. For even though the committee procedure should protect the anonymity of its members, the nature of some criticism makes its source obvious to the VIP. In a study of the use of acupuncture for the prevention of vomiting when narcotics are administered, the VIP intended to use historical controls who had had an anti-emetic drug instead of the acupuncture. This was because he hadn't thought of the acupuncture trial until after the original series. I opposed the use of historical controls believing, for several reasons (see also Pocock, 1983), that they would almost certainly lead to biased results:

1 The rules and forms for the study had not been specified before the historical treatment series.
2 There may have been a change in the type of patient available for treatment.
3 The examining clinician might be more restrictive, either deliberately or subconsciously, in his choice of patients for acupuncture.
4 Those who evaluated the response might be influenced by the knowledge of the different treatment.
5 Patients' attitudes might be different if they knew they were being studied as a special group.
6 It would be difficult to deal with the refusals because even the VIP could not know which of historical controls would have refused.

There were some other reasons for opposing the study but these were the main ones. They led to long arguments, high emotions, and permanent loss of friendship.

Lesson: It is difficult to maintain cool scientific critical detachment.

It may surprise budding consultants that even when researchers acknowledge that statistical advice is sound and of great economic value, the advice is not always taken.

A problem with giant oil rigs is that their welded node plates may eventually fail because of fatigue. A collection of laboratories decided to study the geometries, the welding methods, and the operating conditions that might be controlled to reduce the risk of fatigue failure. They designed a massive experiment costing several million pounds. In effect, each laboratory would study the effect of one factor. This is dubbed 'classical experimentation' which every applied statistician knows to be grossly inefficient when more than a few factors are involved. Such experiments still persist and it should be part of a consultant's training to prepare to counter the classical.

In the case of the oil rigs, somebody did mention the possible inefficiency of their proposed study and I was asked to advise. It was quite difficult because there were 12 factors with from two to five levels each, giving a possible design space of 576000 points except that some of them were barred by constraints. A fractional design could yield all the necessary information at about a tenth of the cost of their original proposal: a saving of millions. However, although they paid for the advice, they did not take it because conservatism among the majority of laboratories forced them to stay with the classical and it was not possible for me to tour them all and to convert each in turn.

It will surely have occurred to you by now that most of the anecdotes of consultations are tales of woe. But these are where the lessons are to be learned. Most research reports are synthetic (as Descartes told us in his *Methods of Discourse* long ago) and are thereby dishonest because they only report success and the authors' brilliance, omitting the failures and wild goose chases. In personal terms, I enjoy being associated with unqualified success. In professional terms we should all be honest and objective with our colleagues, even if it hurts: this is implicit in our professionalism. We gain by recognising and talking about our mistakes and those of our colleagues.

I have been associated closely with only two major multicentre studies. The first was bad and the second was good.

The cot death study (officially 'A multicentre study of post neonatal mortality') aimed, among other things, to describe the epidemiology of children who died between the ages of one week and two years. It involved three years of data collection from eight centres. There were in that time in those centres about 1000 deaths and data were collected for about the same number of controls who were matched for age only. The data were coded onto 15 separate forms for each case, resulting in 1852 variables whose values were punched onto 45 80-column cards. I joined the study as the data collection period was drawing to a close. And found some problems.

There were several laudable achievements from the study. One of the most outstanding was the wide acceptance of home visiting of the bereaved families as a valuable social service as well as a means of investigation. Another was the acceptance of the case conference as a valuable seminar for people from the local health services with a mutual interest. The study also clearly demonstrated the difficulties of getting totally committed and willing participation from all those who must cooperate fully if any venture of this nature is to succeed. Yet I remain sceptical of the study for the following reasons.

1 There was an absence of:
  • plans for data processing
  • a pilot study, and hence of a pilot study analysis and a pilot study report that should have been used in designing the main study

- a full explicit set of hypotheses to be tested, or even any clearly stated hypotheses
- a clear protocol for all data collection and coding, and hence of coding consistency
- any protocol for data analysis with reference to prior stated hypotheses
- any management scheme and progressing system
- training for standardisation of the participants: coders, interviewers, pathologists, health visitors, clinicians

2 The forms were largely useless for coding data for statistical analysis with reference to hypotheses on test. There were far too many variables and yet **no** core set of information that was recorded for every case.

3 There was poor cooperation between clinicians and pathologists with a history of personal clashes.

4 There was a history of major decisions during the course of the study that altered its course.

5 There was bland disregard by some of the pathologists of some of the questions on the forms.

6 There were difficulties in obtaining some population data.

7 There was staggered entry into the study by centres and by districts within the centres.

8 There was continuing discussion, even long after the end of the observation period, about detailed objectives.

These problems led me to make the following brief notes for consideration in the organisation of any future multicentre study.

1 Exact dates for observation should be agreed by all centres and areas within centres to avoid the problem of staggered entry.

2 All jobs should be standardised before the pilot study and again before the main study. This means training all observers and coders.

3 A system should be established for reporting all births in participating districts to the study so that controls can be properly chosen.

4 Similarly, all deaths in the districts should be reported to the study.

5 At least six months should be spent preparing the ground, in explaining in detail the full protocol and the rationale for it to all workers involved.

6 There should be a system for collecting information that does not require any hospital notes to be borrowed by the study centre.

The multicentre study for the evaluation of the medical effects of the seat belt legislation followed these points as closely as their relevance indicated. Patients from car accidents arriving at eight hospitals in England, two in Scotland, one in Wales and four in Northern Ireland in the year before and the year after the introduction of the mandatory wearing of seat belts in the front seats of cars were compared. The study was designed to supplement the national statistics for dead and injured victims of road traffic accidents by showing the effect of the legislation on patients with injuries of different severities and by establishing the relative frequencies of injuries to specific organs before and after legislation. In any major survey such as this, with many variables and cases, it is tempting to indulge in data-dredging: continuing to look through the data to discover effects that have not been thought of beforehand. It is always possible to think of new ways of presenting data as you dredge through it and think of new effects to test. This would certainly lead to spurious inferences and a great waste of time. To avoid this danger, we stated in advance a small number of the most important hypotheses: only 17. Although we did not restrict our statistical analysis to these hypotheses, we confidently claimed statistical significance only when it was based on a prior hypothesis.

The next examples illustrate the value of getting involved with the dirty end of the problem. It is not enough for a statistical consultant to stay seated behind his desk and say: 'State your objectives and I shall design your experiment'; or 'Bring me your data and I shall analyse them.' You must dive into the deep end and discover the real difficulties of experimental management and the real sources of error in data collection. A few practical examples:

A long time ago, when steel was still made in open hearth furnaces, there was a trial to discover the best technique for sampling the liquid metal during the course of steelmaking so that it could be analysed to determine the oxygen content. All the samples couldn't be taken at the same time, because the oxygen content changes. The trial was an inter-laboratory one with representatives from eight laboratories. Each was testing three sampling techniques during both the oxidising period of the steelmaking and the reducing period. We all stood on the deck at the side of the frothing inferno: I with my stopwatch and sampling schedule, shouting orders to the participating chemists. Each stepped forward at my command to thrust his sampler into hell. Some fell off and dissolved (the samplers, not the chemists). Eventually, each solidified sample was sliced into several for repeat oxygen determinations. Finally, there should have been 840 determinations but, because of losses, there were only about 530 and the dataset was very unbalanced. Lack of experience and confidence urged me to seek the advice of a well-known academic. 'Tell them to do it again,' he said.

Staying with oxygen, I planned another inter-laboratory trial for determining the low oxygen content (weight in parts per million) of rolled steel bar. A length of bar was cut into 2-centimetre pieces which were numbered and randomly allocated: nine pieces to each of ten laboratories. It was clear, when the results were returned, that there were some outliers. The question about outliers is: what should be done about them? Should they be rejected because 'they are obviously wrong' as some people argue, or should they be retained because they do represent the natural distribution of errors of determination? My own view is that we should try to find out why the measurements are so far out. Are there errors in measurement, in calibration, in test procedures, or in the source material? Checking back through the randomisation plan, I found that the three furthest outliers had been neighbours in the original bar. This suggested a physical fault, but it could be a coincidence. The three pieces were retrieved from the laboratories. Microscopic examination showed blow holes in the specimens which accounted for their high oxygen determinations. There was now complete physical justification for excluding these three values from the analysis which then showed no exceptional outliers.

Another example of the value of involvement with the practical side of research came from a paediatrician who was developing a technique to determine the gastric emptying and secretion rates of babies. This involved putting measured amounts of water into the stomach, with varying concentrations of dye, at several times, withdrawing small amounts with a tube and syringe and measuring the dye concentrations in the samples. The development was being done using laboratory flasks to simulate the babies' stomachs. The data he gave me behaved very strangely and it was only when I sat in the laboratory with him and actually did some of the procedures myself that I was able to identify several sources of biasing error, such as unmeasured residual liquid in the tube and inadequate graduation of the syringe. But even more important: it helped me to understand more clearly the physical and biological context.

Experience will lead the consultant to be suspicious of data and to be sceptical of claims by the researcher so that it *may* be possible to identify sources of error without moving from the desk.

A metallurgist had a series of 32 casts made in the laboratory's 10-kg furnace to estimate the effects on toughness of several alloying elements. Manganese was specified to be constant throughout at 1.5 per cent but the resulting chemical analysis showed a little variation about that figure. 'But there's no need to include it in the statistical analysis,' he assured me, 'because the variation is completely random and is so small as to have no effect on the mechanical properties.' A simple plot of the data supported his belief, but the next day I asked him if there had been any change in the operating practice after the fifteenth cast. He was surprised but checked the laboratory log book and found that one operator had made the first 15 casts and another the next 17. He was even more surprised when I told him that I had suspected this from the manganese figures, having done a cusum analysis.

Incidentally, this was the same experiment in which regression analysis gave a negative coefficient for vanadium in predicting toughness. 'Ridiculous,' he said. 'Everyone knows that if you want a tougher spanner, you buy one with some vanadium in it. Your computer program is obviously wrong.' I knew it wasn't, and appealed to him to think again about the effect of vanadium at the levels he'd used and in the presence of other alloying elements, considering also the interactive terms that had been estimated. Two days later he returned. 'The results are exactly what should have been expected,' he said. Which again reinforces the point that you should make the consulter document his detailed expectations; that is: record his prior hypotheses.

The counterpart of the consulter who treats you as a technician, is the one who believes you are a brilliant originator. The distribution of responses to an anaesthetic was skewed. I logged them and the distribution became symmetrical. The research anaesthetist thought this was an amazing discovery and wanted to write a joint author paper on the 'original technique'. Fortunately he didn't publish without asking me first. But beware, there are some who will!

I was describing the logrank test for survival data comparisons to a small group. Afterwards a man, whom I didn't know, borrowed my foils for a few minutes. I suspect he photocopied them and am living in dread of reading a paper in which he credits me with originating this 'useful test'.

A final point is a grumble about consulting fees. Many statistical consultants sell their services too cheaply. I suspect this is because, as academics, they already have adequate salaries and see part of the reward for consulting as the source of material for teaching.

Peter had studied the waiting and processing times in an engineering works and had devised a new scheduling system. He didn't know how much to charge for his work.

'How much will your system save them?' I asked.

'At least £20 000 a year.'

'Well charge them £2000.'

Two weeks later he told me he'd been paid £100 and was happy with it.

The 1 per cent rule is often recommended for the independent consultant. Decide on a fair annual salary for yourself then charge a fee of 1 per cent of that for every consulting day. But you may find that the consulter knows Peter.

However, when you are estimating time, never forget Hofstadter's law: 'Every job takes longer than expected, even after taking into account Hofstadter's law.'

Summary of advice for the consultant

- Try to be knowledgeable about all basic science and its applications.
- Play dumb.
- Never give as much information to your clients as you get in return.
- Charge high.
- The best advice is usually simple advice.
- Don't be too willing to help. If your plate is full say 'no' to more.
- Lock the door, switch off the telephone, and refuse to discuss professional problems in the car park or concert hall.
- Work to appointments.
- Don't do emergency *t*-tests for the man who must submit his paper tomorrow.
- Document prior hypotheses.
- Insist on equal status.
- Avoid late involvement.
- Refuse to dredge data.

References

Greenfield, A. A. (1982) The Polymath Consultant. *Proceedings of the First International Conference on Teaching Statistics*, Vol. 2, University of Sheffield, pp. 641–54.

Hyams, L. (1971) The practical psychology of biostatistical consultation. *Biometrics*, **27**, 201–12.

Pocock, S. J. (1983) *Clinical Trials: A Practical Approach*, John Wiley: Chichester.

3

Straight consulting

V. BARNETT

The existence of a strong consulting interest in a university department of statistics (perhaps operating through the provision of a statistical advisory service to colleagues throughout the university) can provide vital lifeblood to that department. It enables the department to keep a high standard within its organisation, in demonstrating that staff are interested, committed and competent in helping others in applied disciplines with the inevitable statistical problems that they face. In return, it provides the statistics department with a most valuable source of material for teaching within a real-life context, for postgraduate students' projects, and indeed for the development of fundamental new research.

Apart from a willingness to undertake such advisory work the consultant statistician does need to have, or to develop, some rather special skills. Some aspects of this matter are discussed in Barnett (1986). I have talked elsewhere of the role of the consultant statistician as 'Jack of all trades – master of one' (Barnett, 1976) and this ubiquitous nature is very much a required characteristic. It is essential that the consulting statistician is able to immerse himself quickly, in an almost chameleon-like manner, in the intricacies of the client's special area (be it heart valves or horse teeth) and to offer the sympathetic communicating manner of the archetypal psychologist in drawing out his client and, at the end of the day, in handing back intelligible, and sometimes not necessarily welcome, conclusions. As I remarked earlier (Barnett, 1976):

> He must be versed in, and capable of handling, the vast array of statistical ideas and methods *per se*. To understand and innovate he needs sound mathematical knowledge.... But...what is of paramount importance is the ability to *apply* statistical knowledge to real problems. In this respect 'the statistician must be a translator and communicator: he needs to understand enough of

other people's disciplines to appreciate their problems. He must express these in statistical terms, in cooperation with the experimentalists, develop and use appropriate tools, and most important, communicate answers in an understandable way. So it would seem that he has a somewhat wider brief than many – as a mathematician-statistician, a computer, a lay philologist (physician, nuclear physicist, you name it) and not least a communicator (this latter facility is not usually regarded as the stock in trade of the scientist!). *All in all he needs to be master of his own statistical trade, but Jack of many others.*

An apologetically phrased request from a client to 'just help me to fit a straight line to these data' will be very familiar to most statistical consultants. Although the request is superficially a modest one it is surprising how often there is a sting in the tail. We shall consider three tales where the opening remark was as above, but the developments were quite distinct. In more than one of the cases there was a need for the production of new and sophisticated methodology to sort out the problems. In all cases the modelling aspect was not straightforward and required a great deal of probing in communication with the client. With at least one of the problems the outcome was quite counter-intuitive, which placed serious demands on the consultant in trying to sell his ideas to the client.

The three problems that we shall consider in the sphere of 'fitting straight lines' arose from the areas of archaeology and medicine, and the methods fortuitously involve us in considering different aspects of the study of a specific field of study: the use of *structural and functional relationships*. This topic of 'regression with errors in both variables' is one in which much still remains to be done from the methodological point of view, although important results have been produced over recent years, sometimes indeed from the very sorts of practical problems described below.

1   Firstly fossils, but not straightforward

As a young and raw member of academic staff my first consulting experience came as a sharp shock. An archaeology PhD student just needed a little help in fitting a straight line to some data on the lengths and breadths of a particular type of (orientatable) fossil shell! The typical data set is illustrated in Figure 1.

Initial relief at just a linear regression problem soon evaporated. There were so many counter indications. The relationship looked non-linear: should we fit a polynomial model? But no: the client gave his assurance that the archaeologists had always believed that lengths (x) and breadths

(y) were *linearly* related. On probing, however, he recalled a 'somewhat irrelevant' piece of information: that there is a fundamental change of shape of the shell with maturity. Perhaps then we were needing a piecewise linear model with an unknown change-point for the slope parameter. In regression terms the model might be

$$y = \begin{cases} \alpha_1 + \beta_1 x + \varepsilon & (x \leqslant x_0) \\ \alpha_2 + \beta_2 x + \varepsilon & (x \geqslant x_0) \end{cases} \tag{1}$$

with $(\alpha_1 - \alpha_2) + x_0(\beta_1 - \beta_2) = 0$ and some appropriately assumed (perhaps normal) distribution for ε to represent the error structure. Even this model with its essentially four-dimensional basic parameter space ($\alpha_1, \alpha_2, \beta_1, \beta_2, x_0$ with a single linking relationship) and single nuisance parameter (the variance of the normal distribution) was by no means an easy one to fit. Indeed at the time (over 20 years ago) relatively little was known about this problem. Some proposals had been made by Sprent (1961) and Quandt (1958). There was soon to be a relative avalanche of work on this topic with major contributions by Hinkley, Quandt, Watts and others (see the annotated bibliography of Shaban, 1980). However, much still remains to be done.

Of course nowadays one also has powerful computer packages (such as GLIM) that can handle the fitting of generalised linear models but the model (1), though apparently simple in structure, is even out of this class.

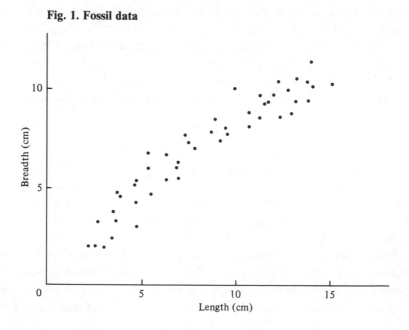

Fig. 1. Fossil data

But the problem does not rest there. The model (1) implies an asymmetry of relationship between x and y, with x error-free and y measured with error ε (or at least an interest in a conditional relationship: the form of y given x). Neither of these was true. Both x and y were error-prone to a similar degree and the archaeologists were not wishing to predict y from x, but to describe the intrinsic (piecewise linear) relationship between them. So we needed a more sophisticated linear model than the linear regression model. One possibility is a *linear functional model* where x and y are assumed to take the form

$$x = u + \varepsilon$$
$$\mathbf{I} \quad y = v + \eta \tag{2}$$

where the error variables ε, η are uncorrelated and possibly normal $N(0, \sigma^2)$ and the relative values of σ_1^2 and σ_2^2 reflect the relative inaccuracies of measurement of the two variables. The model is then completed by declaring that

$$v = \alpha + \beta u \tag{3}$$

and that we observe a specific number (the sample size) of unknown values $u_1, u_2, \ldots, u_n$ of u. Thus we have two basic parameters α, β and $n+2$ nuisance parameters $(\sigma_1^2, \sigma_2^2; u_1, u_2, \ldots, u_n)$; or four and n respectively, if σ_1^2 and σ_2^2 are of importance as basic accuracy measures for the system.

The difficulties in handling the *linear functional model* (2) and (3) are now well known (see, for example, Kendall and Stuart, 1973, Chapter 29), with irresolvable *inconsistency* of estimation when using the maximum likelihood method. (We shall pursue this in more detail later.)

But we are still not at the end of our difficulties with this problem. We really need a *piecewise linear functional model* where (3) is replaced by

$$v = \begin{cases} \alpha_1 + \beta_1 u & (u \leqslant u_0) \\ \\ \alpha_2 + \beta_2 u & (u > u_0) \end{cases} \tag{4}$$

subject to $(\alpha_1 - \alpha_2) + u_0(\beta_1 - \beta_2) = 0$. Even to date no-one seems to have come up with a reasonable method of estimating or testing the parameters in the model (2) and (4). To replace (4) with a non-linear form such as $v = \alpha + \beta u + \gamma u^2$ leads to a somewhat more tractable model (see, for example, Dolby and Lipton, 1972) but it is of course not really appropriate for the job in hand.

So a 'simple' problem from a quarter of a century ago still awaits a full solution: the problem of the day had to be handled with the customary modicum of expedient ad hoccery!

2   Functional protein

We have already remarked that an essential characteristic of a university statistics department is its function as a support service for research and professional work across the spectrum of applied disciplines. Whether this is achieved formally through an organised statistical advisory service, or more casually through the goodwill of individuals in contact with their colleagues in other departments, there is an inevitable flow of benefit in both directions. Few subjects nowadays can advance without the proper use of statistical methods: a do-it-yourself application of computer-based statistical packages is fraught with dangers and we are still far from the realisation of custom-made expert systems (should this ever be a serious prospect)! In reverse, regular contact with assorted practical problems provides the statistician with a flow of research stimuli and invaluable sources of down-to-earth teaching material. The range of topic areas will be rich: perhaps extending from ancient history to zoology. But one thing is sure. If the organisation has a medical school a large proportion of the clients knocking on the statistical consultant's door will inevitably be doctors.

It was from just such a source that my second problem arose. A research physician showed me some data on the protein levels in the blood and urine, respectively, of some patients at different dose levels of a drug. (The

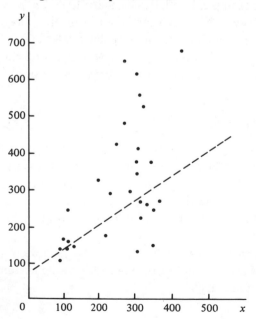

Fig. 2. Data on protein levels

problem is further described in Barnett (1970, 1976).) Typically, such a (suitably transformed) data set appeared as illustrated in Figure 2. Again the request was 'just to fit a straight line': in this case the research interest included demonstrating statistically the long-held medical view that the protein levels in the two sites were intrinsically linearly related.

Once more the complexities began to be apparent. Both variables were subject to errors of measurement: clearly of fairly substantial order to judge from the scatter plot. Furthermore there was no interest in predicting one level from the other. It was a symmetric relationship that was being sought. So again a *linear functional model* had some appeal in the form described by (2) and (3) above. That is, we declare $x = u + \varepsilon$, $y = v + \eta$ with u and v related by $v = \alpha + \beta u$. In the functional model we further specify that u is a *non-random variable*: it takes a specific set of well-defined (but unknown) values $u_1, u_2, ..., u_n$.

Suppose we complete the model by assuming that ε and η are uncorrelated, $N(0, \sigma_1^2)$ and $N(0, \sigma_2^2)$, respectively. This might be so if the u_i correspond to distinct patients. We could try to employ the maximum likelihood approach to estimate the $n + 4$ parameters $u_1, u_2, ..., u_n, \sigma_1^2, \sigma_2^2,$ α and β. The log-likelihood is proportional to

$$L(x, y \mid u, \sigma_1^2, \sigma_2^2, \alpha, \beta) = -n \ln (\sigma_1 \sigma_2)$$
$$-\frac{1}{2} \sum_1^n \frac{(x_1 - u_1)^2}{\sigma_1^2} - \frac{1}{2} \sum_1^n \frac{(y_i - \alpha - \beta u_i)^2}{\sigma_2^2}. \quad (5)$$

One critical likelihood equation takes the form

$$\hat{\sigma}_1^2 = \beta^2 \hat{\sigma}_2^2 \quad (6)$$

with dramatic implications! Here is an extreme form of inconsistency which clearly makes it impossible to draw sensible inferences about the relative values of σ_1^2 / σ_2^2 and β! (Consider how to distinguish different slopes from different ratios of error variance in a scatter diagram.)

So how do we resolve this difficulty? It is not alleviated (as it is for the related *linear structural model* of Section 3) by knowledge of the values of σ_1^2 or σ_2^2 or the ratio $\lambda = \sigma_1^2 / \sigma_2^2$ (and it is in any case hard to imagine realistic situations where such knowledge would exist). For non-normal error structure we can use *ad hoc* estimators based on ratios of cumulants (Geary, 1943) but this is vulnerable to sharp fluctuations if we encounter near-zero values of the cumulants in the denominators. Alternatively, we might be able to simultaneously sample a further *instrumental variable* of appropriate type and exploit the extra information it provides (see, for example, Reiersol, 1945), or use variance components (Tukey, 1951). Special cases of this are the grouping methods of Wald (1940) or Bartlett

(1949) where the data set is divided into non-overlapping subsets and the centroids of the two extreme subsets are joined.

The earliest effort to distinguish formally models and examine anomalies is due to Lindley (1947). For the linear functional model specifically, Solari (1969) analysed the nature of the difficulty and showed that it arose essentially from the fact that whatever the value of β the log-likelihood (5) can be made to approach infinity: no maximum likelihood estimation of β is possible!

Reviews of results for linear functional (and structural) models up to about 1974 are given by Kendall and Stuart (1973, Chapter 29) and Sprent (1969, Chapters 3, 6 and 8). More recent results are briefly described in the Section 4 below.

So what are we to do about the protein level data? In fact, further probing exhibiting extra features of the problem which lead eventually to a full maximum likelihood solution *and to the fitting of the apparently strangely placed 'line of best fit' shown in Figure 2*. How can one possibly justify such a bizarre conclusion to the client? Essentially by a lay version of the following heuristic statistical argument.

We have seen above the difficulty that arises from using the unreplicated linear functional model (*qua* (2) and (3)) – equivalent to assuming each observation to have arisen independently (for example, from different

3   Fig. 3. Behind the protein data (numbers indicate relative weights)

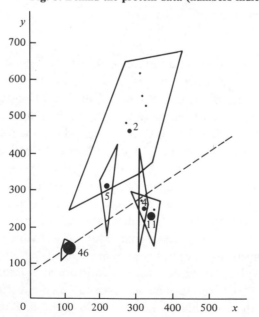

drug levels on separate patients) and with constant error variances, σ_1^2 and σ_2^2. These assumptions, and that of normality, would be difficult to validate in view of the relatively small sample size. However, Figure 2 casts some immediate doubt on the assumed consistency of the error variances: it looks as if they might increase with the true protein levels u and v. Far more important, however, was the almost casual comment by the client that 'of course, the observations do not all arise from different drug levels'. In fact there were only five drug levels administered respectively to groups of 6, 4, 3, 6 and 10 patients. These groups are indicated in Figure 3 from which we do indeed note that the variability seems to increase with the protein level (or correspondingly with changes in the drug level).

So we are really dealing with a linear functional model *with replication* and we might now expect the anomalous behaviour of the maximum likelihood to occur no longer. Dorff and Gurland (1961) have considered various *ad hoc* estimators for the replicated case. But why not try maximum likelihood? This turned out to be quite feasible, even for the rather more complicated heteroscedastic model needed in this problem. Study of several other data sets gave plausibility to a model which said that var(ε) and var(η) both increased with the underlying protein levels u and v, *but in constant proportion to each other*. Specifically, the model now becomes

$$\left.\begin{array}{l} x_{ij} = u_i + \varepsilon_{ij} \\ y_{ij} = v_i + \eta_{ij} \end{array}\right\} \quad (i = 1, 2, ..., p; \quad j = 1, 2, ..., n_i) \tag{7}$$

where i denotes which of the p drug levels is appropriate, and n_i is the number of observations at that drug level. The error model assumes that ε_{ij} and η_{ij} are independent $N(0, \sigma_1^2)$ and $N(0, \kappa\sigma_1^2)$ (with empirical support for σ_1^2 increasing with u_i).

It is now feasible to set up and employ the appropriate more complex form of the likelihood (cf. (5)). It turns out that we cannot obtain explicit (closed-form) expressions for the maximum likelihood estimators of $\alpha, \beta, k, \sigma_1^2, ..., \sigma_p^2, u_1, ..., u_p$, but they exist in well-behaved forms and their asymptotic standard errors can be obtained explicitly. The estimators themselves have to be determined by an iterative numerical procedure in any practical situation and that is what yielded the odd-looking 'line of best fit' in Figure 2.

However, an interesting interpretation of the results is available. In comparison with the likelihood equations for an unreplicated homoscedastic model it is clear that each group of observations (each drug dose level) can be thought of as having an 'equivalent sample size' of n_i/σ_i^2 ($i = 1, 2, ..., p$). Estimating σ_i^2 from the data, this concept of 'equivalent sample sizes' for the groups leads to relative weights of 46, 5, 4, 11 and 2 (cf. sample sizes 6, 4, 3, 6, 10). It is as if we have just five points of vastly different weight to which we fit our 'best straight line'. Figure 3 shows the resulting configuration in terms of which the fitted line now becomes quite plausible.

Of course one would want to do (or to try to do) much more with a problem of this type: examining many data sets and taking specific account of dose levels of the drug. But for illustrative purposes it provides another intriguing example of the complexity of 'simple' problems, of how things are not always what they seem and of how frequently we need to do a bit more research on the way to solving the problem. But who would have expected that the maximum likelihood solution for the replicated linear functional relationship has not been previously published: perhaps someone knows an earlier reference than Barnett (1970)!

3   Heavy breathing

Another medical enquiry (the basis of Barnett, 1969) came from a general practitioner over the 'phone. He wondered if someone at the university could just help a little with the statistical aspects of a research study! It concerned the use of two different types of instrument (a Spirometer: the normal equipment, and a Vitalograph: newly designed) for measuring human lung function. The matter of interest included how much air can be expelled in a single sustained expellation: a crucial measure of 'heavy breathing'! The particular measure we shall discuss is the so-called vital capacity (VC) which expresses the volume of the lungs.

The VC can be measured by either instrument. When it was established that it had in fact been measured on each machine over the *same set of patients*, and that relative calibration and relative accuracy were the matters of interest, warning bells began to sound. The principal features were errors of measurement on each variable, the need to estimate a symmetric rather than a conditional relationship, and the relative values of the two error variances, and *no replication*. Could we be in trouble again?

Once more it was sensible to try a model

$$\left.\begin{array}{l} x = u + \varepsilon \\ y = v + \eta \end{array}\right\} \tag{8}$$

with

$$v = \alpha + \beta u, \tag{9}$$

where x and y are the observed VCs on the Spirometer and Vitalograph, respectively, and ε and η are the measurement errors. Some informal investigation of the data (a typical example with over 70 patients is shown as Figure 4) supported a normal structure. Also, it was more reasonable here to think of the u (and v) values as coming from a population (distribution) of patients rather than taking prescribed fixed values. So u is now thought of as a *random* variable, and (10) and (11) constitute a *linear structural relationship* model. Specifically, the model was completed by assuming that u, ε, η are independent, $N(\mu, \sigma^2)$, $N(0, \sigma_1^2)$ and $N(0, \sigma_2^2)$ respectively.

Until about 1940 there was no distinction between different possible models in the study of 'regression with error in both variables', neither was there any attempt to obtain standard errors of the slope and intercept parameters, or to estimate the basic error components in the two variables (i.e. the error variances, which characterise the scatter about the linear relationship). None the less, there was active interest in the problem from as early as the 1870s (Adcock, 1878; Kummel, 1879), through the work of Pearson (1901) and numerous papers in the statistical and economics journals of the 1920s and 1930s. All the work was concerned with *ad hoc* estimation of the slope and intercept parameters, much of it repetitive (frequently discovering the rather useless prospect of minimising the sum of squares of perpendicular deviations), and many papers mainly expressing critical attitudes towards other writers' proposals. Of course, in the absence of a formulated model (or models), the arguments remained intuitive and personal. Also it was impossible to discuss the wider issues of estimate accuracy and assessment of error variances. As we remarked

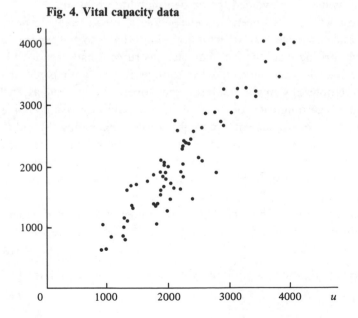

Fig. 4. Vital capacity data

above, it was Lindley (1947) who set the subject on its feet and began to reveal some of the awkward features of the models from the standpoint of parameter estimation.

These are readily illustrated for the linear structural model. Essentially, $(u_1, v_1), (u_2, v_2), \ldots, (u_n, v_n)$ constitutes a random sample from a bivariate normal distribution

$$N(\mu, \alpha + \beta\mu, \sigma_1^2 + \sigma^2, \sigma_2^2 + \beta^2\sigma^2, \beta\sigma^2)$$

The sample means, variances and covariance $\bar{u}, \bar{v}, s_u^2, s_v^2, s_{uv}$ are minimal sufficient and the maximum likelihood estimators of $\alpha, \beta, \mu, \sigma^2, \sigma_1^2$ and σ_2^2 must be functions of them: indeed we should only need to equate the sample moments to their expected values and solve the resulting equations. But there's a big problem! *We have five statistics and six parameters*: so no unique solution. In fact, β is *unidentifiable*, in distinction from σ^2, σ_1^2 and σ_2^2.

It has been pointed out that if we know σ_1^2 or σ_2^2, or σ_1^2/σ_2^2, or σ_1^2 and σ_2^2 then we can obtain the maximum likelihood estimators of the other parameters (see, for example, Madansky, 1959; Barnett, 1967; Birch, 1964). But these are hardly likely prospects! We could also use *ad hoc* methods (cumulants, grouping, instrumental variables, variance components, etc.) as for the linear functional model. (The case of normal error structure causes the most problems!)

So we seem to have encountered a snag with the current problem. If only measurements had been taken twice, on half the number of patients! Such replication would have avoided the unidentifiability problem.

But again we had not been told the whole story in many respects: statistical and circumstantial. It fact it turned out that a sort of replication was present, arising from the fact that each instrument had been used by two operatives *on each patient*, even though it was anticipated that different relationships might hold for each operative as well as each instrument. Regarding each of the four instrument/operative combinations as a separate 'instrument' we might consider extending (8) and (9) in a quite new direction to a set of six simultaneous pairwise linear structural relationships (all pairs from four 'instruments'). The data set with $n = 72$ is shown in Table 1 (from Barnett (1969) where the four variates were denoted y_0, y_1, y_2 and y_3.

The general model is thus one with p instruments used on n patients yielding observations $(x_{1j}, x_{2j}, \ldots, x_{pj})$ $(j = 1, 2, \ldots, n)$ where

$$x_{ij} = u_{ij} + \varepsilon_{ij} \tag{10}$$

The unobserved u_{ij} are observations of random variables U_i related by

$$U_i = \alpha_i + \beta_i U_1 \quad (i = 2, \ldots, p) \tag{11}$$

and the unobserved ε_{ij} come from independent error distributions: $N(0, \sigma_i^2)$. To complete the model we assume

$$U_1 \sim N(\mu, \sigma^2) \tag{14}$$

where $\sim$ means 'is distributed as'. (See Barnett (1969) for more details.)

Table 1. *Readings of vital capacity for the 72 patients on the four instrument/operative combinations*

y_0	y_1	y_2	y_3	y_0	y_1	y_2	y_3
3450	3530	4030	3720	1060	1000	850	600
1310	1320	1610	1600	2000	1800	1270	1700
3820	3720	4150	3700	2280	2280	2380	2350
2110	2880	2740	2520	1940	1800	1670	1580
1860	1420	1540	1690	2580	2700	2850	2110
1940	1780	2020	1800	1400	1440	1680	1480
2360	2260	2430	2350	1260	1100	1000	1030
2880	2920	2650	2860	2320	2420	2360	2360
1980	1720	1800	1660	2000	1940	1980	1980
3120	3180	3250	3040	2400	1900	1470	1740
1760	1630	1390	1200	2880	2980	3240	3140
1480	1760	1700	1640	3420	3150	3200	3200
1840	1660	1400	1650	1000	1130	650	840
3580	3480	3680	3960	1400	1400	1350	1380
1880	2000	2090	2070	1880	1710	1600	1350
2400	2320	2550	2480	1280	1260	1160	1330
2220	2120	2290	2270	3120	3000	3110	3250
2540	2500	2620	1960	3770	3340	3900	3700
920	1200	640	1030	3420	3220	3120	3290
2240	2160	2300	2300	2740	2880	2850	2880
2240	2130	2030	2140	2840	2920	2710	2750
2260	2510	2400	2450	3800	3740	3440	3400
3860	4180	3980	3680	2100	1680	1650	1930
2780	2100	1890	2000	1820	1400	1060	1050
2220	1400	1840	1360	1400	1320	1350	1100
1880	1820	1900	1840	2200	1680	1640	1110
940	960	1060	1000	1940	1900	1820	1270
2480	2220	2150	2150	3260	3200	3250	3270
1660	1780	1760	1800	1960	1940	1890	1920
4040	4180	4000	3770	1320	1260	1140	1000
2540	2560	2080	2250	2840	3060	3650	3510
1780	1700	1390	1200	2060	1840	1720	1780
1280	1300	800	1130	2200	1970	1900	2270
1940	2060	2030	1880	1260	1150	860	1150
1760	2000	1860	1860	3040	2840	2850	2670
2040	1660	1470	1160	2140	2180	2560	2720

Reproduced from: V. D. Barnett, Simultaneous pairwise linear structural relationships. *Biometrics*, **25**, 129–42, 1969. With permission from the Biometric Society.

This richer class of models includes the difficult two-variable case; we need only to set $p = 2$.

We are interested in estimating the parameters α_i, β_i $(i = 2, ..., p)$ and σ_i^2 $(i = 1, 2, ..., p)$ to assess relative calibration, and accuracy, respectively for the different instruments.

We need, however, to be careful about what we mean by calibration. Williams (1969) has drawn the important distinction between (a) absolute calibration and (b) comparative calibration.

In (a) the aim is to interpret readings obtained on a non-standard method in terms of what would have been obtained on some standard method (usually the non-standard method is more economical in some sense). This is merely the prediction problem of classical regression analysis and few problems arise (inverse prediction is a possible difficulty: see Williams (1969)). It is worth noting that this approach does not allow us to talk about the relative accuracy of the two methods.

(b) is typified by the present multiple instrument problem. We want a set of 'conversion formulae' relating results on any pair of instruments: these define the basic relationships between results on the different instruments, uncontaminated by measurement errors, or inherent natural variation. In particular 'common relative calibration' corresponds to an underlying linear relationship of slope 1, through the origin, i.e. $\alpha_1 = 0$; $\beta_1 = 1$ $(i = 1, 2, ..., p)$. Here we have to allow for the measurement errors on all variables, and cut through them to estimate the basic calibration relationships.

To see the advantages of more than one relationship we note that $(X_1, X_2, ..., X_p)$ is p-variate normal with means $\alpha_i + \beta_i \mu$, variances $\beta_i^2 \sigma^2 + \sigma_i^2$ and covariances $\beta_i \beta_j \sigma^2$ $(i, j = 1, 2, ..., p; \alpha_0 \equiv 0; \beta_1 \equiv 1)$. The minimal set of sufficient statistics has dimension $p(p+3)/2$ (the means, variances and covariances) while there are now $3p$ parameters. Consider the case $p = 3$. We have nine sufficient statistics and nine parameters and have only to equate sample moments $\bar{x}_1$ and s_{ij} to their expected values to obtain the maximum likelihood estimators:

$$\hat{\mu} = \bar{x}_1, \ \hat{\sigma}^2 = s_{12} s_{23}/s_{13}$$
$$\hat{\beta}_2 = s_{23}/s_{13}, \ \hat{\beta}_3 = s_{23}/s_{12}, \ \hat{\alpha}_2 = \bar{x}_2 - \hat{\beta}_2 \bar{x}_0, \ \hat{\alpha}_3 = \bar{x}_3 - \hat{\beta}_3 \bar{x}_0$$
$$\sigma_i^2 = s_{ii} - \hat{\beta}_i^2 \hat{\sigma}^2 \quad (i = 1, 2, 3)$$

This is a nicely balanced situation with *no problem of unidentifiability*. Furthermore, asymptotic standard errors are easily obtained (Barnett, 1969).

Consider general values of p. The numbers of sufficient statistics and parameters are as follows.

p	$p(p+3)/2$ sufficient statistics	$3p$ parameters	Condition
2	5	6	unidentifiable
3	9	9	'balanced'
4	14	12	
5	20	15	
.	.	.	identifiable
.	.	.	
.	.	.	

So unidentifiability is unique to the case $p = 2$! For $p > 2$ there are no formal obstacles to applying maximum likelihood, although explicit expressions for estimators cannot be obtained beyond the balanced case $p = 3$. But this presents no serious problem. Iterative solutions are feasible (or even approximate solutions based on subsets of three out of the p relationships) and asymptotic standard errors are obtainable. For details see Barnett (1969).

The special nature of the unidentifiability problem for the $p = 2$ case is further highlighted by considering another extension of the model: to all *multilinear structural relationships* between any m out of p variables $(2 \leqslant m \leqslant p)$. In this wider class again $m = p = 2$ stands out as a sore thumb – *and only one other case*, that of a single bilinear structural relationship $(p = 3, m = 3)$ (see Barnett, 1979).

Another intriguing prospect arises. We have referred to the use of instrumental variables to try to overcome the unidentifiability problem for a single linear structural relationship. Earlier proposals of this type suffered from being essentially impractical. The required conditions were seldom likely to be satisfied. But why not exploit the simplicity of the balanced case $(m = 2, p = 3)$ and merely seek as instrumental variable *any one further measure* which is linearly structurally related to the two variables of principle interest? This is often quite readily achieved.

So to return to our original lung function problem, we can summarise the situation as follows. An *apparently* simple problem of linear relationship *seemed* to suffer from an insurmountable problem of anomalous behaviour of the estimators. What *seemed* an added complexity (additional linear structurally related variables) in fact yielded a sort of replication and *enabled a solution to be found*. In the process it opened up a new area of research on this topic and finally shed light on another way forward on the original anomalous situation of a single linear structural relationship.

4   **Have we finished?**

Is there nothing more to be said about the study of functional and structural relationships? On the contrary, it is still a rich research area. Much has been achieved over the last 10 years or so, including different representations (for example, through factor analysis: see Theobald and Mallinson, 1978), ultrastructural models which extend but imbed both the functional and structural cases (see, for example, Dolby, 1976), non-linear functional models (see, for example, Dolby and Lipton, 1972: there is perhaps a fundamental osbtacle to an equivalent non-linear *structural* model) and Bayesian methods (see, for example, Lindley and El-Sayyad, 1968). So it is still a very fertile area, and much more useful work can be expected both in employing such models to solve important practical problems and in using such problems to stimulate important new methodology.

References

Adcock, R. J. (1878) A problem in least squares. *The Analyst*, **5**, 53–4.

Barnett, V. (1967) A note on linear structural relationships when both residual variances are known. *Biometrika*, **54**, 670–1.

Barnett, V. (1969) Simultaneous pairwise linear structural relationships. *Biometrics*, **25**, 129–42.

Barnett, V. (1970). Fitting straight lines – the linear functional relationship with replicated observations. *Applied Statistics*, **19**, 135–44.

Barnett, V. (1976) The statistician: Jack of all trades, master of one? *The Statistician*, **25**, 261–79.

Barnett, V. (1979) Unidentifiability and multilinear structural relationships. *Bull.Int.Statist.Inst.*, **48** (4), 49–52.

Barnett, V. (1986) Two degrees under, or how to avoid catching cold in the oil industry. *To appear* in: Holmes, P. (ed.) *Saving Money with Statistics*.

Bartlett, M. S. (1949) Fitting a straight line when both variables are subject to error. *Biometrics*, **5**, 207–12.

Birch, M. W. (1964) A note on maximum likelihood estimation of a linear structural relationship. *J.Am.Statist.Ass.*, **59**, 1175–8.

Dolby, G. R. (1976) The ultrastructural model: a synthesis of the functional and structural relation. *Biometrika*, **63**, 39–50.

Dolby, G. R. and Lipton, S. (1972) Maximum likelihood estimation of the general non-linear functional relationship with replicated observations and correlated errors. *Biometrika*, **59**, 121–30.

Dorff, M. and Gurland, J. (1961) Estimation of the parameters of a linear functional relationship. *J.R.Statist.Soc.* B, **23**, 160–70.

Geary, R. C. (1943) Relations between statistics: the general and the sampling problem when the samples are large. *Proc.R.Irish.Acad.*, **49**, 177–96.

Kendall, M. G. and Stuart, A. (1973) *The Advanced Theory of Statistics*, vol. 2, 3rd edition, Griffin, London.

Kummel, C. H. (1879) Reduction of observed equations which contain more than one observed quantity. *The Analyst*, **6**, 97–105.

Lindley, D. V. (1947) Regression lines and the linear functional relationship. *J.R.Stat.Soc.Supp.* **9**, 219–44.

Lindley, D. V. and El-Sayyad, G. M. (1968) The Bayesian estimation of a linear functional relationship. *J.R.Statist.Soc.* B, **30**, 190–202.

Madansky, A. (1959) The fitting of straight lines when both variables are subject to error. *J.Am.Statist.Ass.*, **54**, 173–205.

Pearson, K. (1901) On lines and planes of closest fit to systems of points in space. *Phil.Mag.*, **2**, 559–72.

Quandt, R. E. (1958) The estimation of the parameters of a linear regression system obeying two separate regimes. *J.Am.Statist.Ass.* **53**, 873–80.

Reiersol, O. (1945) Confluence analysis by means of instrumental sets of variables. *Arkiv for Matematik, Astronomi och Fysik*, **32**, 1–119.

Shaban, S. A. (1980) Change point problem and two-phase regression: an annotated bibliography. *Int.Statist.Rev.*, **48**, 83–93.

Solari, M. E. (1969) The 'maximum likelihood solution' of the problem of estimating a linear functional relationship. *J.R.Statist.Soc.* B, **31**, 372–5.

Sprent, P. (1961) Some hypotheses concerning two-phase regression lines. *Biometrics*, **17**, 634–45.

Sprent, P. (1969) *Models in Regression and Related Topics*, Methuen, London.

Theobald, C. M. and Mallinson, J. R. (1978) Comparative calibration, linear structural relationships and congenic measurements. *Biometrics*, **34**, 39–45.

Tukey, J. W. (1951) Components in regression. *Biometrics*, **7**, 33–70.

Wald, A. (1940) Fitting of straight lines if both variables are subject to error. *Ann.Math.Statist.*, **11**, 284–300.

Williams, E. J. (1969) Regression methods in calibration problems. *Bull.Int.Statist.Inst.*, **43**(1), 17–28.

4

A two-period crossover trial

D. CLAYTON AND M. HILLS

1   Introduction

In a crossover clinical trial with two treatment periods each patient starts by receiving one of the treatments in the first period and then *crosses over* to receive the other treatment during the second period. Referring to the two treatments as A and B, there are thus two groups of patients: those starting with A and crossing over to B (group 1) and those starting with B and crossing over to A (group 2). Equal numbers of patients are allocated at random to the two groups but drop-outs can cause the final numbers to be unequal. Trials like these frequently figure in consultancy work because interpreting the results from them can be difficult, both for the physician and the statistician. On this occasion the client, a pharmaceutical company, asked us to analyse the results from a crossover trial in which the response to treatment took the form of a continuous 24-hour electrocardiogram.

The trial was one of a number, carried out under varying conditions, in which a new drug was compared with a standard treatment. We knew, therefore, that our analysis would not be viewed in isolation but as part of a whole body of evidence, and that it might well form part of a submission to a regulatory authority such as the FDA (Food and Drugs Authority, USA). This meant that any analysis we carried out would have to meet the following requirements:

(i)   the statistical method used should be simple and easily understood by someone reading our report along with those from other trials;

(ii) the analysis should provide an estimate of the size of the treatment difference;

(iii) any assumptions made in the analysis should be justified.

We tried to meet these requirements by using a simple graphical technique combined with a distribution-free estimate of the amount by which one distribution is shifted in location relative to another. The theoretical basis for the analysis of two-period crossover trials has been reviewed by Hills and Armitage (1979). The discussion below has much in common with their treatment of the problem but we would hope that the graphical extension we propose below will help to clarify the argument for non-statisticians.

In a typical crossover trial the response to treatment for each patient is observed twice, once for each period. We shall be concerned with a quantitative response to treatment and shall regard the two responses as observed values of the variables (Y_1, Y_2), where Y_1 refers to the response for period 1 and Y_2 to the response for period 2. The expected values of (Y_1, Y_2) will be denoted by $(\theta_{A1}, \theta_{B2})$ for group 1 and $(\theta_{B1}, \theta_{A2})$ for group 2. This notation indicates which treatment is used during each period for the two groups.

The shift in location of the distribution of Y_1, from group 2 to group 1 is equal to $\theta_{A1} - \theta_{B1}$, the treatment difference for period 1. Similarly the shift in location for the distribution of Y_2, in the other direction, from group 1 to group 2, is $\theta_{A2} - \theta_{B2}$, the treatment difference for period 2. The hope is that these two treatment differences will be the same, so that a pooled estimate based on both periods can be used. If they are not, then the treatment difference in period 1 is the only one with any clinical meaning so that we would only use the results from period 1.

A simple way of checking that the treatment differences are the same in both periods, and of obtaining a pooled estimate if they are, is to change from the variables (Y_1, Y_2) to new variables (Z_1, Z_2) where $Z_1 = \frac{1}{2}(Y_1 - Y_2)$, $Z_2 = \frac{1}{2}(Y_1 + Y_2)$.

From the definition of Z_1

$$E(Z_1 \mid \text{group 1}) = \tfrac{1}{2}(\theta_{A1} - \theta_{B2})$$
$$E(Z_1 \mid \text{group 2}) = \tfrac{1}{2}(\theta_{B1} - \theta_{A2})$$

so that the shift in location from group 2 to group 1 for Z_1 is

$$\tfrac{1}{2}\{(\theta_{A1} - \theta_{B2}) - (\theta_{B1} - \theta_{A2})\} = \tfrac{1}{2}\{(\theta_{A1} - \theta_{B1}) + (\theta_{A2} - \theta_{B2})\} \tag{1}$$

Similarly, from the definition of Z_2,

$$E(Z_2 \mid \text{group 1}) = \tfrac{1}{2}(\theta_{A1} + \theta_{B2})$$
$$E(Z_2 \mid \text{group 2}) = \tfrac{1}{2}(\theta_{B1} + \theta_{A2})$$

so that the shift in location from group 2 to group 1 for Z_2 is

$$\tfrac{1}{2}\{(\theta_{A1} + \theta_{B2}) - (\theta_{B1} + \theta_{A2})\} = \tfrac{1}{2}\{(\theta_{A1} - \theta_{B1}) - (\theta_{A2} - \theta_{B2})\}. \tag{2}$$

It follows that if there is no shift in location for Z_2 then from equation (2) the treatment difference is the same in both periods and from equation (1) an estimate of its value may be obtained from the shift in location for

Z_1. This method is particularly useful where a pooled distribution-free estimate of the treatment difference is required.

A plot of Z_1 against Z_2 for all patients in the trial, distinguishing between those in group 1 and those in group 2, thus provides a simple and convenient summary of the results: if the shift between groups in a horizontal direction is small then the shift between groups in a vertical direction is an estimate of the treatment difference (see Figure 1). It is also possible to assess from the plot whether the distributions of Z_1 and Z_2 within two groups differ in more complex ways than a shift in location. If they do, then this suggests that the observed values of Y_1 and Y_2 should be transformed in some way (usually by taking logs) so that the treatment effects are more nearly constant over the patients on the transformed scale, and hence that the distribution of the resulting values of Z_1 differ only by a shift in location. The assumption that two distributions differ only by a shift in location is important when making a distribution-free estimate of the shift and it is usually worth looking at the two cumulative relative frequency curves for Z_1 (and for Z_2), as well as at Figure 1, in order to check that the assumption is valid.

When the distributions of Z_2 for the two groups provide strong evidence of a shift, and therefore of different treatment effects in the two periods, there is often considerable interest in why this might have occurred. One possibility is 'carry-over', where the effect of the treatment in period 1 carries over into period 2. This could account for unequal treatment differences in the two periods if the carry-over effect from treatment A differs from the carry-over effect from treatment B. Another possibility in that the response changes with time so that it tends to be much lower

Fig. 1. Idealised Z_1/Z_2 plot

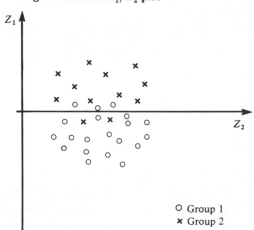

O Group 1
× Group 2

(say) in the second period than the first, with the result that there is less room for the treatment differences to be expressed in the second period, causing it to be smaller than in the first period.

One difficulty with the two-period crossover trial is that the test for equal treatment differences in the two periods based on Z_2 may not be very sensitive as the variability in Z_2 $(= \frac{1}{2}(Y_1 + Y_2))$ includes variability in the overall level of response *between* subjects, which may be rather large.

This has led some investigators to use three periods, still with two treatments, and to use the treatments in the order A–B–B for group 1 and B–A–A for group 2. The difference between the two treatments is then assessed using $Z_1 = Y_1 - \frac{1}{2}(Y_2 + Y_3)$, where Y_1, Y_2, Y_3 refer to the responses for periods 1, 2 and 3 respectively, and the carry-over is assessed by using $Z_2 = Y_2 - Y_3$. Both Z_1 and Z_2 are now influenced only by variation *within* subjects, a feature which makes the *three*-period *two*-treatment design an attractive proposition where practical considerations allow its use. A more theoretical discussion of three-period designs is given by Kershner and Federer (1981) who show that, in the presence of residual effects, the ABB + BAA design is four times as efficient as the alternative ABA + BAB design.

Another improved design incorporates a placebo 'washout' period between the two periods of active drug therapy. This was used in the trial reported below and, as we shall see, in addition to minimising pharmacological carry-over effects, also allows a more powerful test for differential carry-over. However, use of placebos is often not ethically acceptable and in such cases the three-period design is to be recommended.

2   The trial

The trial we consider was a comparison of two anti-arrhythmic drugs in the treatment of ventricular premature complexes (VPCs) (Kjekshus, Bathen, Orning and Storstein, 1984). VPCs are disturbances of normal heart rhythm, and are visible on an electrocardiographic trace.

Figure 2a illustrates the ECG trace corresponding to a normal heart beat and shows the normal QRS complex flanked by the P and T waves. In a VPC, the QRS complex is abnormal in shape and occurs abnormally early (Figure 2b). While occasional VPCs occur in the normal heart, they occur much more frequently in patients suffering from various forms of heart disease.

A number of drugs have been shown to be effective in suppressing VPCs, but the condition is a chronic one and no permanent cure is effected. Thus, for the comparison of two such drugs, a crossover design may be considered.

This trial compared a new compound (Flecanide acetate) with an established anti-arrythmic drug (Disopyramide) in a two-period crossover design. Each active drug was given for a two-week period. Although patients with frequent VPCs are at increased risk of sudden death, there is no firm evidence that drugs which suppress VPCs bring a corresponding benefit in survival. There is no clinical consensus that such patients should be routinely treated with anti-arrythmic agents and in these circumstances it was felt to be ethically acceptable to incorporate 'washout' periods in the design, during which patients would be treated with placebo. There was a one-week washout period on entry to the trial before the first active period, a further one after the first active period and before the second, and a final one on completion of the second active period. Thus, each patient was in the trial for a total of seven weeks.

The purpose of placebo washout periods is to minimise any residual effects of treatments given to patients before their entry to the trial, and to minimise 'carry-over' of effects from the first active period into the second. Since patients are randomised for the two treatment order groups, the former effect will not cause a bias but a washout period might improve the efficiency of the trial by more closely controlling the initial conditions.

Fig. 2. ECGs (a) the normal electrocardiogram, (b) a premature aberrant complex (marked X)

(a)

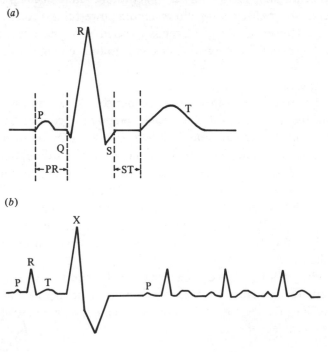

(b)

Carry-over between the two active periods is, however, a more serious problem and, as we shall see, data collected during the washout periods can be useful evidence for its occurrence. In the present study, the elimination of both drugs from plasma is relatively rapid, the half-life being of the order of one day. Thus, a one-week washout period is adequate to ensure that no trace of the first active drug remains during the second active period.

The frequency of arrythmic complexes was determined by recording the ECG for the last 24 hours of each week of the trial. The measurements were made as far as possible without disturbing the patient's normal way of life by using a portable medical data recorder capable of recording 24

Table 1. *Ventricular premature complexes per 24 hours*

Week	1	2	3	4	5	6	7
	(placebo)	(active)		(placebo)	(active)		(placebo)
Group A		*Flecanide*			*Disopyramide*		
	0	2	5	5	6	6	0
	x	2352	1557	6229	1527	2554	7603
	1180	4	144	5274	1	12	1162
	153	x	22	41	0	1	0
	15008	771	222	23950	12759	11735	20040
	2611	638	x	700	1201	14272	2792
	1610	0	0	5976	0	2	154
	1961	6	0	94	2075	68	x
	63346	21262	2317	50579	49617	15932	83862
	x	366	217	26300	13452	19943	37682
	173	98	13	268	74	17	x
	2873	0	6	473	12	30	1005
	20702	0	0	2761	5	173	6570
	160	11	913	x	466	2064	x
Group B		*Disopyramide*			*Flecanide*		
	579	1364	3640	866	453	47	x
	674	17	138	649	1	7	1468
	16868	4663	41	6016	336	2108	7066
	0	143	1	9	0	1	26
	9534	*	*	45	12	740	1
	3356	5050	4912	1014	12284	7648	1829
	x	1167	758	1133	372	x	2019
	4448	2682	x	2057	8216	1756	312
	9930	13025	10714	x	x	x	158
	37274	19718	25881	18925	0	14	17747
	624	8	2	820	0	0	271

x data missing because of unanalysable tapes.
* data missing because of discontinuation of drug.

hours of ECG on a C120 cassette. Each recording concerns some 1 000 000 heart beats, and was processed by a high-speed computer analyser. A number of different types of abnormality are counted by the analyser, but we shall consider here only the data concerning total frequency of VPCs; this data is shown in Table 1.

Each row of the table shows the total VPCs recorded for one patient in each of the seven 24-hour recordings. Although 30 patients were originally randomised, the table gives data for only 25 patients. Two patients died during the study, and three had to be withdrawn owing to their developing serious complications. Of the remaining 25, plasma analysis showed that two patients had not taken the drug (Disopyramide) and the readings during that active period were discounted. Ambulatory electrocardiographic recording in free-living subjects presents some practical difficulties, and it is perhaps not surprising that some of the tapes proved to be unanalysable.

3   The response

We now consider the problems we encounter when relating the real data of Table 1 to the theoretical background discussed in the opening section.

The first problem is that each patient was, ideally, measured on seven occasions, while our theory calls for only two response measures, Y_1 and Y_2. It might be argued that condensing the data from seven to two observations per patient should be avoided since some information must be lost. However, the analysis of the total data is difficult, since we can be almost sure that the assumption of independent errors is invalid in a sequence of observations in time. The problem for the consultant is to find a method of analysis which provides accurate point and interval estimates of the relative efficacy of the treatments; makes a relatively efficient use of the data; and is comprehensible to the scientific community to whom the work is addressed. This compromise is not always an easy one and we should not worry too much about modest losses of efficiency if the other two aims are achieved. It is worth remembering that there are many practical, ethical and clinical reasons for patients being seen regularly during a trial, and measurements are often taken because the patient is there rather than for any strong scientific purpose. Thus, we should not be afraid to lump together some measurements and discard others.

In this trial we average the two readings in each active period to form Y_1 and Y_2. In so doing we may also solve another problem, that of missing observations: when both measurements are available we take their average, but if either is missing we simply take the measurement which is present. When both measurements are missing, then the responses Y_1 or Y_2 cannot be estimated and hence neither can the contrasts Z_1 and Z_2. In such cases no information can be salvaged without strong assumptions and we must omit that patient from this analysis.

Table 2(a) shows these simplified data. Purists would argue that the fact that some measurements are based upon 48 hours recording and some are only 24 hours gives them different precisions and this information should be carried forward into the later analysis. This is, however, a case in which the small gain in efficiency does not justify the increased complexity of the analysis.

A more serious concern is whether this procedure, which amounts to replacing missing observations by the average of the available observations, might cause bias. This would occur if the process causing loss of data were

Table 2. *The transformed data*

| | (a) VPCs per 24 hours | | (b) $\log_e(\text{VPCs}+\tfrac{1}{2})$ | |
| | | | Period 1 | Period 2 |
First drug	Period 1	Period 2	(Y_1)	(Y_2)
Flecamide	3.5	6	1.39	1.87
	1954.5	2040.5	7.58	7.62
	74	6.5	4.31	1.95
	22	0.5	3.11	0.00
	496.5	12247	6.21	9.41
	638	7736.5	6.46	8.95
	0	1	−0.69	0.41
	3	1071.5	1.25	6.98
	11789.5	32774.5	9.37	10.40
	291.5	16697.5	5.68	9.72
	55.5	45.5	4.03	3.83
	3	21	1.25	3.07
	0	89	−0.69	4.49
	426	1265	6.14	7.14
Disopyramide	2493	250	7.82	5.52
	77.5	4	4.36	1.50
	2352	1222	7.76	7.11
	72	1	4.28	0.41
	*	376	*	5.93
	4981	9966	8.51	9.21
	962.5	372	6.87	5.92
	2682	4986	7.89	8.51
	11896.5	*	9.83	*
	22799.5	7	10.03	2.01
	5	0	1.70	−0.69

* data missing because of discontinuation of drug.

related either to the level of response, or to the treatment itself. Here, however, the pattern of missing values indicates no such problems.

The third problem which these data present is their extreme variability; the counts range over several orders of magnitude. In such cases it is usually necessary to transform the data before proceeding to further analysis. The procedure of data transformation is one of the most difficult problems in statistical consultancy as the rationale which underlies it is almost universally misunderstood by non-statisticians. The most common misconception is that the main reason for transformation is to make the data more closely normally distributed, and this encourages a widespread mistrust of transformation, too often regarded as a statistician's device to force data into his techniques rather than to further scientific understanding.

In fact, we have no need to assume a normal distribution in the analysis to be presented below. Indeed in small studies where there is scarcely enough data to examine normal assumptions adequately, normal theory methods should be avoided if possible in favour of distribution-free methods. The reason for transformation of the observations is to find an appropriate scale on which the assumptions of our model are at least an adequate approximation to reality. The important assumptions are (i) additivity of the effects of period and treatment, and (ii) additivity of these systematic effects and the other extraneous random influences. These assumptions are necessary for the scientific purpose of separately estimating these components in the total response of the patient. Without the first additivity assumption it is not possible to estimate separately an index of relative efficacy of treatments from this data. The second assumption holds that the *variance* of the response is not affected systematically by treatment or period. This is sometimes referred to as the assumption of 'homoskedasticity', but is more simply regarded as a special kind of additivity assumption. As we have indicated, the assumptions of the analysis can be examined by plotting the contrast Z_1 against Z_2.

Clearly, however, there is no hope of the additivity assumptions holding true on the original scale of measurement (VPCs per 24 hours), because the data vary over several orders of magnitude and those patients tending to show greater VPC frequencies also appear to show a greater lability of response. In such cases the natural first choice is to log-transform the observations. This must improve matters and still leads to a readily interpretable analysis since a linear contrast is expressible in terms of a ratio of the measurements on the original scale.

We encounter a difficulty when taking logs because some of the counts are zero. The best practical method for dealing with this is to use the empirical log transformation log $(n+\frac{1}{2})$ in place of $\log n$ where n is the observed frequency. A formal justification of adding $\frac{1}{2}$ to the frequency is given by Snedecor and Cochran (1967) but for our purposes the precise choice of $\frac{1}{2}$ as a 'handicap' to each frequency is not too critical since we later use rank methods. The empirical log-transformed VPC counts are shown in Table 2(b) and by simple inspection can be seen to exhibit a more regular pattern than those in Table 1.

Technically, if we are to choose the transformation to optimise the fit of the additive model, we should adopt formal criteria and take account of the estimation of the transformation in determining the sampling variances and significance of our estimates. For example, we might find the transformation from the power-law family of parametric transformations investigated by Box and Cox (1964) in which a single parameter picks a particular transformation (including both logarithms and identity transformations). Recent work generalises this to allow non-parametric estimation of the transformation. However, in practice, we are probably justified in ignoring these refinements. Where possible we work with the original scale but when, as here, the measurements are positive values which vary widely, then the log scale is natural. We need only abandon these natural scales if there is evidence for major deviation from the model assumptions.

At this point it is relevant to comment upon a procedure which is becoming widespread in the analysis of pharmaceutical trials and is potentially seriously misleading. This involves transformation of the original observations to ranks; i.e. replacing each observation by its position in the ranking of all observations. These rank values are then analysed as if they were genuine observations. The original of this procedure is a paper by Conover and Inman (1981) which advocates this strategy as a 'bridge between parametric and non-parametric statistics'. While, in certain areas, this procedure reproduces simple well-known non-parametric methods, in general it is not well founded. In the analysis of the crossover trial the reason for transformation is for the separation of period and treatment effects and we transform for additivity. There is no reason why the transformation to ranks should be a good transformation to additivity; indeed there is every reason why this scale, which is bounded at both ends, is likely to prove inappropriate for linear modelling.

A correct rank-invariant approach involves non-parametric estimation of the scale transformation to additivity. However, such methods are in their infancy and lead to some difficulties of interpretation. Where possible, therefore, we recommend analysis on one of the natural scales, but adopt distribution-free methods for testing and estimating the treatment effects on this scale. In the next section we show that this strategy is successful here, despite the initially daunting distribution of observations.

4   The analysis

The observed values of Y_1 and Y_2 are shown in Table 2. They are transformed counts for each treatment period using the transformation

$$Y = \log \text{ (number of VPCs} + \tfrac{1}{2})$$

The values of $Z_1 = \tfrac{1}{2}(Y_1 - Y_2)$ are plotted against the values of $Z_2 = \tfrac{1}{2}(Y_1 + Y_2)$ in Figure 3. Inspection of this figure suggests that there is no shift between groups in the horizontal (Z_2) direction, so the treatment difference is the same in both periods, but there is a shift between groups in the vertical (Z_1) direction. This impression is further corroborated by the cumulative relative frequency curves which are used to display the observed distributions of Z_1 and Z_2 in the two groups. Figure 4 shows the two distributions of Z_2 to be close, while Figure 5 shows the two distributions of Z_1 to have the same shape but different locations.

Formal tests of significance and estimations of shift based on the Wilcoxon rank sum test are shown in Table 3. Note that when using this method the estimated shift is found from the median of the $14 \times 9 = 126$ differences between an observation from group 1 and an observation from group 2, not from the difference between the medians. The 95% confidence intervals for this shift are also distribution-free and represent the maximum and minimum value, δ say, which may be added to *all* the observations

Fig. 3. Z_1/Z_2 plot for the data of Table 2

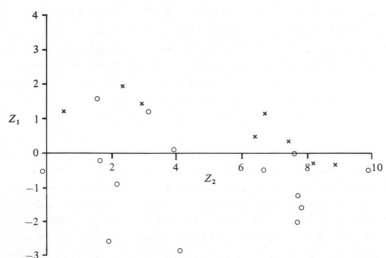

of Z_1 (or Z_2) in group 2 so that a Wilcoxon test for a location shift is not significant at the 5% level.

The formal test of significance for Z_2 shows that there is no evidence that the treatment difference changes from period 1 to period 2 and the pooled estimate for the treatment difference (from Z_1) is -1.71 with confidence limits -2.84 to -0.56. This estimate is on the transformed scale and corresponds to a value for the ratio of the frequency of VPCs on Flecanide to that on Disopyramide of $e^{-1.71} = 0.81$ with confidence

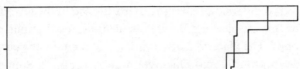

Fig. 4. Cumulative relative frequency curves for Z_2

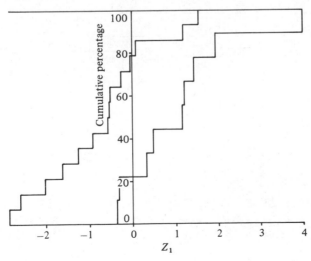

Fig. 5. Cumulative relative frequency curves for Z_1

limits from $e^{-2.84} = 0.06$ to $e^{-0.56} = 0.57$. Thus the new drug reduces the frequency of VPCs to something between 6% and 57% of what it would be on the existing treatment.

The plot of Z_1 against Z_2 in Figure 3 shows that, as expected, the variability in Z_2 within groups is larger than that of Z_1: the range within groups for Z_2 is about twice that for Z_1. Thus the test for equal treatment differences in the two periods (based on Z_2) is rather insensitive by comparison with the test and estimate based on Z_1. Fortunately, in this case, we have data from the placebo periods and can use this to back up the evidence from Z_2. The most likely reason for unequal treatment effects in the two periods is differential carry-over, mentioned in the introduction. We can test for this by checking the difference (placebo 3 minus placebo 2) between groups. The placebo 2 observation will contain any carry-over from Flecanide in group 1 and Disopyramide in group 2. Similarly the

Table 3. *Results of Wilcoxon tests: crossover analysis*

	Z_1	Z_2
Median for group 1 ($n = 14$)	-0.53	4.02
Median for group 2 ($n = 9$)	1.15	6.39
Estimated shift from group 2		
to group 1	-1.71	-0.68
95.3% confidence limits for	-2.84	-4.3
the shift	-0.56	1.6
Significance of the shift (P)	0.005	0.64

Fig. 6. Cumulative relative frequency curves for placebo differences

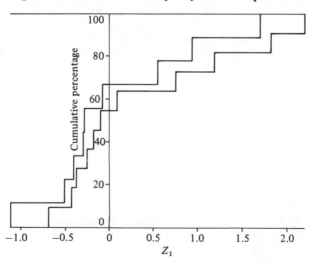

placebo 3 observation, will contain any carry-over from Disopyramide in group 1 and from Flecanide in group 2. Thus the contrast, $\frac{1}{2}$ (placebo 3 minus placebo 2) should show a shift from group 1 to group 2 equal to the difference in the carry-over effects of the drugs. The two distributions are shown in Figure 6 and a Wilcoxon rank sum test confirms that there is no significant shift between them ($P = 0.45$). This test is based on the distributions of *differences* between placebo periods within the same subject and these have about the same variability as values of Z_1. (The ranges for the placebo differences and for Z_1, within groups, are both about 3.)

Had there been any suspicion of differential carry-over we would have had no choice but to restrict our analysis to the data gathered in the first period, and here we show this analysis for the sake of comparison. One possibility is simply to estimate the treatment effect by the location shift of the distribution of Y_1 from group 2 to group 1. Theoretically, however, one might expect that a more efficient analysis should be in terms of the changes ($Y_1 - Y_0$), from the first placebo baseline observations, Y_0. These analyses are shown in Table 4.

The analysis of Y_1 shows a significant treatment difference but, as expected, at a less stringent level, the width of the 95% confidence interval for the treatment difference being considerably larger than for the corresponding interval based upon Z_1. Interestingly, the analysis of ($Y_1 - Y_0$) is considerably less successful, partly because of a loss of a further three observations owing to unanalysable ECG tapes in the placebo period. These analyses demonstrate clearly the gain in efficiency achievable with crossover design when its assumptions are met.

Table 4. *Results of Wilcoxon tests: parallel group analysis*

	Y_1	$Y_1 - Y_0$
Number of observations		
Group 1	14	12
Group 2	10	9
Median		
Group 1	4.17	−2.34
Group 2	7.79	−0.49
Estimated shift, group 2		
to group 1	−2.95	−2.45
95% confidence limits	−5.62	−6.51
for the shift	−0.41	+0.47
Significance of shift	0.015	0.08

5   The report

When writing the report of the analysis of a clinical trial it is important to bear in mind that the report must serve at least two different purposes. It is a report to the company developing the new drug and also to the investigators who carried out the scientific work. These are different audiences with slightly different concerns and expertise.

Let us consider, first, the needs of the pharmaceutical company. No drug will ever be granted a marketing licence without extensive trials and any one trial will form only one small part of a comprehensive programme. Such programmes are rather loosely subdivided in phases and the premarketing phases are from phase 1 (studies in healthy volunteers) to phase 3 (final controlled clinical trials under realistic conditions). The economy and efficiency of crossover trials makes them a very attractive choice of method at phases 1 and 2 when, for example, different dose levels might be contemplated, but the fact that each period of active treatment is of necessity rather short makes them less attractive for phase 3 testing. Thus, the place of crossover trials is predominantly in phases 1 and 2 and here the prime aim is to obtain reliable estimates of efficacy of the drug in different groups of patients, in different doses, and compared with different alternative therapies. Ultimately, reports concerning all trials in the testing programme must be brought together to form a submission to drug regulatory bodies.

Statistical significance testing has little to contribute to this process. What is required is a report which clearly sets out a model for drug efficacy and provides point and interval estimates. The report must also examine critically the model assumptions. The US drug regulatory authority explicitly requires that evidence from crossover trials be accompanied by *internal* evidence that carry-over effects were not present.

In an ideal world the needs of the clinical investigators would be little different from those of the pharmaceutical company but in reality a style of presentation of statistics in medical journals has developed which makes crossover trials rather hard to present in an acceptable manner. The 'unobtrusive' style for statistics in medical journals consists of presenting simple tabulations or graphs of the data, liberally decorated with statistical significance levels. It would be quite commonplace to use the data of Table 1 presented either graphically or in table form in terms of conventional summary statistics, such as mean and standard deviation, for each of the seven measurements in each group. This satisfies the clinician's desire to see response profiles for the treatment groups rather in the same manner as the progress of two individual patients might be displayed, but is irrelevant to the proper analysis of the data. Clinical trials are comparative studies which are concerned with differential response of individuals to different treatments. In the crossover trial where treatments are compared within individuals, the analysis must be primarily concerned with *differences* between responses rather than with their absolute values. However, there continues to be some resistance among clinicians to tabulation of the differences, Z_1. We hope that the Z_1/Z_2 plot of Figure 3 will become acceptable for displaying the results of crossover trials. It contains all the relevant data for estimation of effects and for criticisms of the model in a readily understood form.

Another problem of presentation arises because of the use of the logarithmic transformation so that, for example, Z_1 represents change on a scale which may be unfamiliar to a clinician. With the use of the log transformation, the additive model on which the analysis depends may be expressed as a multiplicative model on the original scale. By taking antilogs, additive treatment effects on the log transferred counts may be expressed as multiplicative effects and these are much more readily understood.

The analysis of the crossover trial relies heavily upon a linear model, as do many analyses in modern applied statistics. The familiarity which statisticians develop towards these models is very definitely not shared by researchers in other disciplines, and communication can be difficult. This is particularly the case when, as here, transformation of data is necessary. However, with care and with simple graphical aids, analysis may be performed, and presented to the client in a comprehensible manner.

References
Box, G. E. P. and Cox, D. R. (1964) An analysis of transformations.
J.R.Statist.Soc. B **26**, 211–52.
Conover, W. J. and Inman, R. L. (1981) Rank transformation as a bridge between parametric and non-parametric statistics. *The American Statistician*, **35**, 124–9.
Hills, M. and Armitage, P. (1979) The two-period crossover clinical trial. *Br.J.Clin.Pharmacol.*, **8**, 7–20.
Kershner, R. P. and Federer, W. T. (1981) Two treatment crossover designs for estimating a variety of effects. *J.Am.Statist.Ass.*, **76**, 612–19.
Kjekshus, J., Bathen, J., Orning, O. M. and Storstein, L. (1984) A double-blind crossover comparison of flecanide acetate and disopyramide phosphate in the treatment of ventricular premature complexes. *Am.J.Cardiol.*, **53**, 72B–8B.
Snedecor, G. W. and Cochran, W. G. (1967) *Statistical Methods* (Sixth edition), Ames: Iowa State University Press.

5

Consultancy in a medical school, illustrated by a clinical trial for treatment of primary biliary cirrhosis

D. G. COOK AND S. J. POCOCK

The role of academic statisticians in a university medical school can be quite varied. In addition to teaching commitments, their research activities may be classified into three broad areas:

(1) availability as a *statistical consultant* to advise and provide technical assistance for the statistical aspects of any worthwhile research projects;

(2) to undertake more in-depth *scientific collaboration* on a limited number of research projects that contain a major statistical element; such projects will often be primarily motivated within the statistician's own department, but may sometimes arise from contacts outside the medical school;

(3) to undertake *methodological research* in medical statistics to develop new approaches to the design and analysis of medical investigations.

The balance of time that each statistician devotes to these three areas will depend on the nature of their appointment, their own particular aptitudes, the opportunities that arise in each area and the general policy of the department and medical school in which they work.

In our case, the Department of Clinical Epidemiology and General Practice at the Royal Free Hospital School of Medicine in London has active research programmes in epidemiology and clinical trials so that the principal research effort is directed towards scientific collaboration and methodological research. This requires applied statisticians to expand their activities into non-statistical areas, by acquiring the necessary knowledge of the medical and epidemiological issues relating to the collaborative research. That is, statisticians have to become *project-oriented* whereby their efforts will be directed towards advancing medical knowledge. Any statistical details, particularly the use of complex mathematical or computational procedures, must be genuinely relevant to the medical problem and must not impair the ability to communicate findings to non-statistical colleagues. In general we feel that the skill to communicate beyond their own profession needs much greater emphasis in the training of graduate statisticians.

At the Royal Free, we have had to decide how to organise a statistical consultancy service to the medical school. There are limited resources available for this general consultancy because of the more in-depth collaborative and methodological research mentioned above. Therefore our emphasis is on: (a) an advisory service to all 'reasonable people' seeking statistical help; (b) more detailed technical assistance to a limited number of projects and; (c) occasionally a more substantial scientific collaboration (an example of which forms the bulk of this chapter).

The advisory service is mainly about short-term one-off consultations whereby some fundamental issue in study design or statistical analysis needs sorting out. Advice is often straightforward, indeed some statisticians are not enamoured by the trivial level (statistically) of many such consultations. However, in practice the statistician can derive considerable stimulus from such exposure to a wide variety of medical problems. Providing advice can be a very time-consuming business and hence we have found this best organised by having a statistician available one afternoon every week in a *statistics clinic* at which 'customers' can book appointments. Of course, off-the-cuff enquiries are still dealt with on occasions but this rationalisation of statistical advice has generally improved its handling both from the point of view of statisticians and 'customers'.

In general, the lack of resources prevents us from providing a more detailed technical service for analysing other people's data. Most projects wish to remain self-sufficient in that respect anyway and routine 'number crunching' is an unrewarding activity for an experienced statistical scientist. Nevertheless, we do undertake some limited technical help, usually in analysis but occasionally in design aspects of the problem. The analysis issues we tackle ourselves usually go beyond the knowledge of the client; problems such as analysis of survival data, multiple regression or analysis of variance.

The art of the matter is then to present these analyses and their interpretation back to the investigator in a form that they can clearly understand. In practice, we have sometimes arranged such analyses to be undertaken by a graduate student for an hourly fee.

The most rewarding aspect of statistical consultancy arises when communication between the statistician and the researcher moves into a more *long-term scientific collaboration*: the medical investigator wishes to explore fully the statistical aspects of their research and the statistician wishes to be absorbed into the scientific purpose of the whole project so that their role is no longer peripheral. Because of the time commitment this requires, a statistician can only achieve such a role in a limited number of projects at one time and these will often be confined to just one or two medical departments.

At the Royal Free School of Medicine we have built up such a relationship with some members of the Academic Department of Medicine. To illustrate how such collaborative statistical activities can unfold, the remainder of this chapter describes the development of our contribution to one particular clinical trial.

1   Initial contact

In 1980 we were approached by Dr Owen Epstein, a clinician from the Medical Unit at the Royal Free Hospital, concerning the analysis of a placebo-controlled randomised clinical trial of the drug D-Penicillamine (D-P) for the treatment of primary biliary cirrhosis (PBC), a rare liver disease. The trial was a single-centre trial and had been recruiting patients at the Royal Free since 1973. When the clinician approached us there had been two deaths out of 55 patients on treatment compared with eight out of 32 on placebo.

Although based on a small number of deaths, the difference looked highly significant, which is precisely what had prompted the clinician to contact us. In fact there was no provision in the trial protocol for when the trial results should be analysed. Since no predetermined stopping rule existed we faced a problem: by analysing the trial at a time when the clinician felt that a difference had become apparent, a bias in favour of finding a statistically significant difference would exist. On the other hand, it was unethical to continue the trial if the treatment was beneficial. In the event a compromise solution was reached: it was decided that a more definitive analysis would be undertaken six months later.

2   Some comments on the trial design and clinical background

PBC is a chronic progressive liver disease of unknown cause, which is thought to be an essentially auto-immune disease. There is no effective treatment. From the onset of symptoms the median survival is around 13 years. However, in many cases a correct diagnosis may not be made for several years after symptoms appear.

Since the Royal Free is a referral centre for patients from all over Britain and abroad, one could expect that survival of those in the trial would be shorter. D-P is an immuno-suppressive drug which is also effective in lowering liver copper. Since PBC is often complicated by retention of copper in the liver this suggested D-P as a possible therapeutic agent.

D-P is fairly toxic and side effects occur frequently. These require careful monitoring and often lead to withdrawal of treatment. Thus, although the trial was designed to be double blind with a placebo control, the patient's physician often became aware of the treatment, because of side effects. However, this should not be a major problem, since the outcome measure is patient survival. In 1972 when the trial was being set up, a statistician was consulted and it was decided that, since so many withdrawals were expected from the D-P group, randomisation should be unbalanced: two-thirds of patients were entered on D-P and one-third on placebo. According to the current consensus, which strongly recommends analysis of clinical trials on an 'intention to treat' basis (i.e. patients are included for analysis in the group to which they were randomised, irrespective of whether treatment has stopped or not), this imbalance was perhaps unnecessary though it did enable the physicians to gain greater experience of giving D-P treatment. Randomisation was carried out via the hospital pharmacist who supplied blinded drug packages according to a sequence of treatment assignments (D-P or placebo) in random permuted blocks.

3   The first analysis and publication

By the time of the first formal analysis, 5 (9%) of 55 D-P treated patients and 10 (31%) of 32 placebo patients had died. A check was made to see that no large imbalances in prognostic factors existed between the treated and control groups at entry. All such differences were small and were such as to suggest that the treated group had a slightly worse prognosis at entry. This point is taken up again in the second analysis which we carried out in 1984 described below. The main analysis was a comparison of the two survival curves (Figure 1) using the logrank test. Since no patients with early-stage disease (stages 1 and 2) had died, the analysis was restricted to those with more advanced disease at entry (stages 3 and 4). It is worth noting that one of the reasons for preferring the logrank test to fitting Cox regression models (which would have allowed adjustment for prognostic factors) was that such models are far more difficult to communicate to physicians. At this stage of involvement in the trial, statistical complexity appeared counterproductive. Since the difference was so highly significant ($P < 0.01$ using the logrank test) the clinical investigator and ourselves decided it was unethical not to publish the findings. These appeared in *The Lancet* in June 1981 (Epstein *et al.*, 1981) with the following conclusion:

> The excellent prognosis of patients with PBC in its early histological stages, and the failure of D-P to prevent histological progression from early to late stages, suggests that D-P treatment should not be given to patients with PBC in the early (stage 1 or 2) histological phase of the disease. D-P treatment is recommended to patients once liver biopsy has demonstrated histological results typical of later stage 3 or 4 PBC.

4   After publication: continuing the trial?

A decision now had to be taken on whether or not the trial should continue. The decision had to be made against the background of two other even smaller trials, both of which had shown no significant survival benefit on D-P treatment (Matloff, Alpert, Resnick and Kaplan, 1982; Triger, Manifold, Cloke and Underwood, 1980).

In the event, entry to the trial was stopped, but those patients already entered were continued in the trial on their current treatment. This

Fig. 1. Survival curves from first analysis in patients with late histological stage (stages 3 and 4) PBC. Below horizontal axis, n = number of penicillamine-treated patients and placebo-treated patients with late-stage histology at risk of dying at each time period

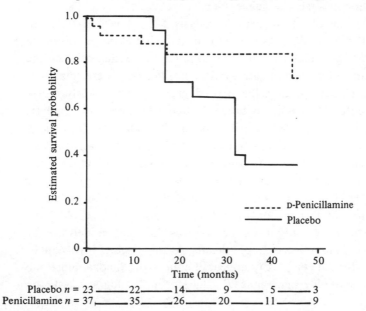

Placebo n = 23 ——22——14—— 9 —— 5 —— 3
Penicillamine n = 37 ——35——26—— 20 ——11—— 9

somewhat curious decision reflects the ethical and scientific dilemma posed by the apparently conflicting findings of three rather small trials. It seemed particularly important to continue follow-up of patients already in the trial so that the potential advantage of D-P treatment could be more definitively resolved. Such a policy might seem to conflict with the *Lancet* paper's conclusion, but in practice it is extremely difficult to achieve the right balance of ethical and scientific judgement, particularly when the disease is rare and further trials are unlikely to take place.

5   Second analysis 1984

In January 1984 we performed a further and more detailed analysis. The trial now had 98 patients (11 more patients entered between first analysis and stopping entry) with a median follow-up of 48 months. There were 18 (30%) deaths of 61 penicillamine patients compared with 16 (43%) deaths of 37 placebo patients. Survival still favoured the penicillamine group (Figure 2), but the difference was no longer statistically significant ($P = 0.09$ using a logrank test). Restricting analysis to patients in the late stage (stages 3 and 4) made the difference less significant, since three patients in the early stage (stages 1 and 2) had now died: one on treatment and two on placebo.

Had our first analysis merely reflected the bias we had feared? That is, did the timing of publication occur when the treatment difference was

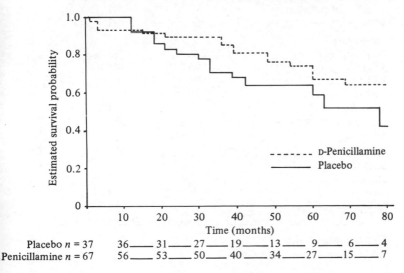

Fig. 2. Survival curves from second analysis in all patients. Below horizontal axis, n = number of penicillamine-treated patients and placebo-treated patients at risk of dying at each time period

somewhat inflated? Possibly, yet the survival curves still favoured the treated group even if the difference was no longer significant at the 5% level: $P = 0.09$ is not that far removed.

6   Prognostic factors

We decided to look in more detail at the data, in particular at the importance of the following variables measured on entry: age; liver copper (Cu); stage (based on a liver biopsy, and scored from 1 to 4); hepatomegaly (a binary variable indicating an enlarged liver); granulomas (a binary variable indicating the presence of granulation tissue in the liver); and several biochemical markers in the blood: albumin, bilirubin, aspartate transaminase (AST), alkaline phosphatase (AP) and IgM (a measure of immune status). Some of the variables were highly skewed and were log transformed before analysis.

Table 1 shows that the two groups were broadly similar at entry into the trial, though what differences do exist suggest that the D-P group had a somewhat worse prognosis. Contrary to popular belief, formal significance testing is inappropriate here since it has no sensible basis in a randomised trial and fails to indicate the clinical importance of any differences in prognostic factors. One variable in 20 should have a treatment difference which is statistically significant at the 5% level by

Table 1. *Comparison of treatment groups at entry to trial*

	Placebo	Penicillamine
General		
Number of patients	37	61
Mean age (years) (range 30–77)	55	53
Laboratory		
Bilirubin* (g/l)	27	30
Albumin (g/l)	42	43
AST* (units/l)	40	51
AP* (KAU/l)	54	65
IgM* (g/l)	5.2	4.8
Cu* (μg/g dry weight)	162	183
Histologic		
Stage 1	4 ⎫ 30%	4 ⎫ 33%
Stage 2	7 ⎭	16 ⎭
Stage 3	17 46%	22 36%
Stage 4	9 24%	19 31%
Hepatomegaly present	32%	28%

* Geometric means.
KAU, King Armstrong units.

definition, while non-significant differences between groups at entry may still affect the treatment comparison if the variable in question is strongly prognostic (Altman, 1985). As an example, and in the light of subsequent results, it is worth noting that the *t*-value for the difference in baseline bilirubin levels between treatment groups is 0.9.

Table 2 gives the *p*-values for each variable's association with patient survival, using a univariate proportional hazards model. This indicates the powerful influence that some variables (for example, bilirubin) have on patient survival.

We used a stepwise procedure based on the proportional hazard model to determine which combination of prognostic factors was most suitable for predicting patient survival. Variables were entered sequentially if they were significant at the 5% level, while any variables in the model which became non-significant were dropped. We ran this both starting from the null model and from the model including all variables.

Since no deaths occurred in those without hepatomegaly, entering it into the Cox regression model amounted to restricting our analysis to those 78 patients with hepatomegaly. No estimate of the relative hazard is possible. In practice we found that the model obtained by restricting analysis to those without hepatomegaly excluded stage, but was otherwise identical. We ended up with a model which included albumin, (log)AST, (log)AP and stage (or hepatomegaly). Hepatomegaly and stage are highly confounded.

The omission of bilirubin was a surprise, since it is generally accepted

Table 2. *Univariate analyses*

Variable	*t*-statistic for hazard coefficient	*p*-value
(log)bilirubin	5.6	0.0001
Albumin	−4.9	0.0001
(log)AST	4.4	0.0001
Hepatomegaly	*	0.0001
Stage (1 & 2 v. 3 & 4)	3.1	0.0017
(log)Cu	3.2	0.001
Granulomas	1.9	0.06
Age	1.3	0.21
(log)IgM	−1.0	0.32
(log)AP	−0.1	0.90
Treatment	1.7	0.09

* Calculated using logrank test since no patient without hepatomegaly died.

as the single most important clinical prognostic indicator. However, (log) bilirubin is highly correlated with (log)AST ($r = 0.06$) and the choice is not so clear-cut that we would expect AST to be preferred in all data sets. In fact bilirubin was the single most important indicator, but became non-significant once both albumin and AST were included. What we can say is that albumin is an important prognostic factor independent of bilirubin or its correlates. The inclusion of AP was also unexpected, since it was nowhere near significant in the univariate analysis. It was decided to retain it in the model since our clinical collaborators had noticed that before death AP tended to fall in some patients which is in agreement with the negative coefficient. In fact there was some indication that the effect was greater in patients with later stage disease at entry.

7   Goodness of fit

There are essentially two assumptions inherent in our model: (*i*) the assumption of proportionality and (*ii*) the additivity of different factors in contributing to the log hazard. Many goodness of fit tests are now available (see, for example, Kay, 1984). However, it is unrealistic to expect them to have much power in such a small data set. We merely dichotomised each variable on the basis of its distribution and looked at the cumulative hazards on a univariate basis to ensure they did not cross.

8   Treatment difference after allowing for prognostic factors

Table 3 gives the result of fitting a proportional hazards model including both prognostic variables and treatment as a dichotomous variable. The *P*-value for treatment is now 0.04. However, caution is required here just as we were cautious of dismissing a treatment effect when looked at on its own. The 95% confidence limits for the treatment

Table 3. *Significant prognostic variables and their regression coefficients in the final Cox regression model*

Variable	Scoring	Regression coefficient	Standard error	P
Albumin	g/l	−0.25	0.057	0.0001
AST	$\log_e$ (value in units/l)	2.69	0.526	0.0001
AP	$\log_e$ (value in units/l)	−1.14	0.394	0.0039
Stage	1 & 2 = 0 3 & 4 = 1	1.53	0.626	0.0146
Treatment	Placebo = 0 D-P = 1	−0.78	0.377	0.0394

coefficient are wide $(-1.53, -0.02)$. That is, we are 95% certain that the effect of D-P is to multiply the hazard by between 0.22 and 0.98. Such uncertainty reflects the small number of deaths in the trial and its consequent lack of power for detecting even relatively large treatment effects.

9   Instability of the model

The choice of prognostic variables influenced both the magnitude of the estimated treatment effect and its significance. If AP was dropped from the final model then the P-value for treatment was only 0.09. A model including only bilirubin gave a P-value of 0.03 for treatment, while if albumin was added P became 0.22. This instability was entirely due to changes in the estimated treatment coefficient, the standard error remaining stable. Such instability is disconcerting, but not unusual given the size of the trial.

Since the P-value and confidence limits for the treatment are so dependent on the choice of covariates, it is desirable that the selection of variables be taken into account when calculating them. One method of achieving this is to use bootstrap methods (Gong, 1982). That is, one generates new data sets by randomly sampling with replacement from the observed data set. One might thus generate 500 data sets, each with 98 observations, select the covariates for each data set as before, and then estimate the treatment effect. The resulting distribution of the 500 estimated treatment coefficients could then be used to make statements about the probable magnitude of the treatment effect. We are currently looking into this possibility, but for the present we return to our more conventional analysis.

10   Interpreting the proportional hazards model

Whatever the analysis, the results must be understood by the clinician. When research findings are being reported the analysis should not only be valid; its presentation should be understood by the non-statistical reader. This was why no proportional hazard models were fitted in the first paper: they would have obscured the simple message concerning a treatment effect. If any imbalance in prognostic factors had existed in favour of the D-P group then we might have acted differently.

In the reanalysis such simplicity was insufficient. However, even to the statistician the hazard coefficients in Table 3 are of limited value; to the clinician they are gobbledegook. To communicate the results of such an analysis, further interpretation is required.

Proportional hazard models are appearing increasingly in the medical literature. After one paper (Schlichting *et al.*, 1983) an editorial 'surviving proportional hazards' (Elashoff, 1983) appeared in *Hepatology* illustrating the need felt by clinicians to comprehend the analysis of such data and the difficulty they have in doing so.

While familiarity is leading to general acceptance of such papers, this does not always mean greater comprehension. Statisticians need to be more than just consultants in such studies. They need to be fully involved in design, analysis and interpretation, but most of all they must communicate a complex analysis to an audience with limited numerical skills. Only by communicating successfully will statisticians maintain respect; if they do not succeed then, as survival packages become more accessible and better documented, there lurks the danger of analyses without the statistician.

11   Adjusted survival curves

A survival curve is far better understood by a clinician than is the hazard function. In the absence of covariates it is a straightforward matter to calculate the Kaplan–Meier survival curves for each treatment separately (Figure 2). When covariates are involved it seems logical to calculate estimated survival curves for the placebo and D-Penicillamine groups for central values of the covariates. We do not present such curves here since they differ little from Figure 2, but an example in which the difference is quite dramatic is to be found in Christensen *et al.* (1985). Instead of basing the estimated curves on the final model, an alternative approach is available both within the SAS procedure PHGLM and within BMDP2L. It is possible to calculate the log-likelihood function separately for the placebo and treated groups; the two components are then added to yield the overall likelihood which is maximised for parameter estimation. It is assumed that the regression parameters are the same within each stratum, though no assumption is made concerning the relationship of treatment and the hazard. The underlying survival curves can then be calculated for each treatment group adjusting all covariates to central values. The advantage of this stratified approach is that proportionality of hazards is not assumed between treatment groups.

12   Conclusions concerning the trial

In retrospect the PBC trial was always too small to answer satisfactorily the question of treatment efficacy, which is not to say that nothing was gained from it. (James (1985) reviews the results from seven controlled trials of D-P for the treatment of PBC, including the Royal Free trial.) It seems possible that D-Penicillamine is of some benefit in the treatment of PBC, but we are still some way short of the certainty we would like and we do not know whether the benefit is minimal or a major clinical advance. Clearly the lack of a stopping rule in the protocol created problems, but the major problem was one of sample size.

Suppose we were planning the trial from scratch and we performed a sample-size calculation. If the five-year survival without treatment was 40% and we were interested in detecting a reduction of 25% in five-year mortality from 40% to 30%, then to stand a 90% chance of detecting such a difference at the 5% significance level we would require 470 patients on *each* treatment followed for five years. No trial of PBC has ever approached this size and it would clearly require a multi-centre trial at an international level to do so.

It is worth making the point that in a situation where a clinician approaches a statistician to calculate the sample size for a prospective clinical trial, the clinician's estimate of a hoped-for reduction in mortality is likely to be highly optimistic. Trials planned on the basis of such estimates frequently turn out too small.

A further point is that PBC is a disease with a long natural history. It is therefore unlikely that patients with early-stage disease at entry will die during the course of the trial. Thus they add little to the power of the trial. Similarly there may be patients with advanced disease at entry who will die before any treatment can have an effect. Sample-size calculations should, but often do not, take account of this natural history of the disease.

We are currently using the data from the D-Penicillamine trial to look carefully at the natural history of PBC. Each patient's bilirubin was measured every three months from their entry into the trial. We wish to use this to predict better when a patient becomes at high risk of dying. This is relevant, for example, in deciding when patients are candidates for liver transplants. From the clinical point of view, what is required is a prediction that a patient is likely to die, within two years say, unless transplanted, but this prediction must be made before the patient becomes too ill for a transplant to be carried out. In this context one obvious extension of the proportional hazards model is to make the coefficient for bilirubin time-dependent and these issues are being further explored.

Lastly, we are faced with the problem of how to communicate the updated findings of this trial to a general medical audience. The estimated magnitude of treatment difference is somewhat reduced compared with the initial publication in 1981 but it is generally difficult to get such updated information into a major medical journal. This illustrates the general problem of how statistically significant findings based on small numbers may lead to investigators (and statisticians) conveying more enthusiastic conclusions regarding therapeutic progress than are really justified. Further emphasis needs to be given to such 'publication bias' and perhaps medical journals should offer opportunities for authors to provide short updates of their trial data to keep the medical public fully informed.

13   Some general remarks

We chose to present the above experiences with a clinical trial, not because it was an earth-shattering piece of research but because it illustrates several of the issues that statisticians are likely to encounter when consulting. First, we were only consulted at the analysis stage, though admittedly another statistician had been superficially involved in the initial design. One 'parrot cry' from statisticians is to be consulted when the project is being planned. Although this is obviously desirable, if every investigator were to do so we would all be grossly overwhelmed with design problems. Thus, when investigators do make first contact with data in hand we do not usually find it rewarding to adopt a militant enquiry about why they had not come earlier. To build up an effective relationship it is best to tackle the investigator's immediate problems rather than lecture him or her about the past. Similarly, in choosing an appropriate analysis strategy one should consider the fact that simple techniques presented clearly can enhance communication more readily than sophisticated methods which the investigator may have difficulty in comprehending. As the relationship progresses, the interested investigator will become more receptive to advanced techniques and then it is possible to work together on how to report such methods in the medical literature.

On the whole relatively few statistical consultations can be expected to lead to a close and continuing scientific collaboration or to an interesting opportunity for methodological research. However, by adopting an effective style of communication with each investigator, the applied statisticians can greatly enhance their own opportunity for a stimulating and varied career in medical research.

References

Altman, D. G. (1985) Comparability of randomised groups. *The Statistician*, **34**, 125–36.

Christensen, E., Neuberger, J., Crowe, J., Altman, D. G., Popper, H., Portmann, B., Doniach, D., Renek, L., Tygstrup, N. and Williams, R. (1985) Beneficial effect of azathioprine and prediction of prognosis in primary biliary cirrhosis. Final results of an international trial. *Gastroenterology*, **89**, 1084–91.

Elashoff, J. D. (1983) Surviving proportional hazards. *Hepatology*, **3**, 1031–5.

Epstein, O., Jain, S., Lee, R. G., Cook, D. G., Boss, A. M., Schuer, P. J. and Sherlock, S. (1981) D-Penicillamine treatment improves survival in primary biliary cirrhosis. *Lancet* **1**, 1275–7.

Gong, G. (1982) Some ideas on using the bootstrap in assessing model variability. In Heiner, K. W., Sacher, R. S. and Wilkinson, J. W. (eds.) *Computer Science and Statistics: Proceedings of the 14th Symposium on the Interface*. New York, Springer Verlag, pp. 169–73.

James, O. F. W (1985) D-Penicillamine for primary biliary cirrhosis. *Gut*, **26**, 109–13.

Kay, R. (1984) Goodness of fit methods for the proportional hazards regression model: a review. *Rev. Epidem. Sante Publ.*, **32**, 185–98.

Matloff, D. S., Alpert, E., Resnick, R. W. and Kaplan, M. M. (1982) A prospective trial of D-Penicillamine in primary biliary cirrhosis. *New Engl. J. Med.*, **306**, 319–25.

Schlichting, P., Christensen, E., Anderesen, P. K., Fauerholdt, L., Juhl, E., Poulsen, H. and Tygstrup, N. (1983) Prognostic factors in cirrhosis identified by Cox's regression model. *Hepatology*, **3**, 889–95.

Triger, D., Manifold, I. H., Cloke, P. and Underwood, J. C. B. (1980) D-Penicillamine in primary biliary cirrhosis: two year results. [Abstract] *Gut*, **21**, A919–20.

6

The analysis of response latencies

G. DUNN

1 Introduction

As a statistical consultant and teacher in a multi-disciplinary research establishment such as the Institute of Psychiatry one is faced with many types of statistical problem. One also has to learn to cope with a wide range of statistical abilities and/or experience in the clients. Not only may the client not have any proficiency in the use of statistics and computers, but there are also wide variations in the overall approach to research in the different disciplines. Anthropologists, psychiatrists, psychologists and biochemists have very different views about the role of research and the way it should be carried out. Clearly a statistical consultant has an obligation to provide advice about good experimental or survey design together with a description and explanation of the appropriate methods of analysis. Perhaps the exact nature of the advice will depend on the background of the client and the context in which the client is carrying out the research. But, if the consultation is taking place in an academic setting, such as a postgraduate medical school, what else should be provided? Should one expect the *client* to learn to use a computer and carry out the analysis? Yes. Ideally, I think that the consultant statistician should be regarded as a teacher rather than as a 'boffin' to be used to help generate the right answers. Given this view, should the analysis be kept simple (but perhaps not as 'perfect' as it might have been if carried out by the expert) or should it be relatively sophisticated? Psychophysiological experiments, for example, often involve relatively complex designs for the collection of repeated measurements on relatively few human volunteers or patients. Should the resulting data be analysed using the repeated-measures options from a complex program such as MANOVA from the SPSSX package (SPSS Inc., 1983) or should the client be encouraged to consider a few relevant contrasts and analyse them using simple *t*-tests or the analysis of variance? The simple approach is more readily followed and understood by the statistically naive client and his or her peers. The complete analysis, on the other hand, conveys more information even though it may go well beyond the research workers' *essential* needs.

In this chapter I present a fairly detailed picture of an example of the handling of a particular type of repeated measurement. These measurements are of latencies or response times involving the possibilities of censoring and competing risks. They were generated by a psychological experiment involving the recall of pleasant or unpleasant memories. As a statistical consultant my primary concern was to teach the client (a psychiatrist, Dr Din Master) how to look at and analyse data resulting from this *and* future experiments. The material has also been used as an introduction to survival data and the corresponding statistical models for other research workers in the psychiatric and behavioural sciences (see Dunn and Master, 1982; Everitt and Dunn, 1983, chapter 10).

2   Background: the basic experiment

In the basic experiment, subjects are presented with stimulus words on a visual display unit and are asked to think of pleasant and unpleasant memories associated with these words (Lishman, 1974). The time between the presentation of the stimulus and the subsequent memory recall is recorded and, in this chapter, will be referred to as a recall latency or recall time. It has been shown that, on average, subjects who are not depressed recall pleasant memories faster (or more easily) than unpleasant ones. Depressed subjects, however, tend to recall unpleasant memories faster than pleasant ones (Lloyd and Lishman, 1975). Normal subjects who have been made to feel sad using various experimental techniques can also be shown to recall unpleasant memories more easily then pleasant ones (Teasdale and Fogarty, 1979).

The details of the experiments on memory recall times are as follows. Each subject is seated in front of a CBM model 4032 desk-top computer and the procedure explained using practice word lists. Subjects are presented with a sequence of words (such as 'black' or 'table') on the computer's screen, one at a time, each word being preceded by an instruction to try to remember a pleasant or unpleasant experience or event associated with the word. Requests for pleasant and unpleasant memories alternated in the sequence. In the full experiment different groups of subjects can be given different word sequences.

Pressing the keyboard bar produces the stimulus word, along with the reminder of the category of recall required. Successful recall of a memory is indicated by the subject again pressing the bar. The computer records the time between the first and second pressings of the bar. If a recall cannot be achieved within 15 seconds of the first bar-press, the stimulus word is replaced by an instruction to clear the mind and, when ready, to proceed with the next word. In this case the recall time is recorded as a censored observation of 15 seconds. Following each successful recall, the stimulus word is replaced by an instruction to rate the memory as mildly, moderately or very pleasant (or unpleasant).

A period of practice is given until the experimenter is sure that the procedure is fully understood by the subject. In total, 48 stimulus words are presented, half requiring pleasant and half requiring unpleasant recalls. To summarise, a single subject produces 48 recall times, 24 for pleasant memories and 24 for unpleasant ones. Each recall time can be censored at 15 seconds, and quite often is. Finally, for each recall there

Table 1. *Memory recall times (t) for one male subject* (Taken from Dunn and Master, 1982).

Pleasant	Unpleasant	Observed survival function
1.07	1.45	0.96
1.11	1.67	0.92
1.22	1.90	0.88
1.42	2.02	0.83
1.63	2.32	0.79
1.98	2.35	0.75
2.12	2.43	0.71
2.32	2.47	0.67
2.56	2.57	0.63
2.70	3.33	0.58
2.93	3.87	0.54
2.97	4.33	0.50
3.03	5.35	0.46
3.15	5.72	0.42
3.22	6.48	0.38
3.42	6.90	0.33
4.63	8.68	0.29
4.70	9.47	0.25
5.55	10.00	0.21
6.17	10.93	0.17
7.78	15+	0.13/0.17*
11.55	15+	0.08/0.17*
15+	15+	0.08/0.17*
15+	15+	0.08/0.17*

* The first number refers to pleasant memories, the second to unpleasant ones.

is *competition* between mildly, moderately or very pleasant (or unpleasant) memories. For the preliminary phase of the description of the analysis, however, the intensity of the effect associated with the memories is ignored. An example of data collected for a single male subject (me) is shown in Table 1. The observations have been ranked and given an associated value of the survival function to aid interpretation.

3   Preliminary analysis

The aim of the preliminary analysis was two-fold: first, to look at the responses from one or two subjects to get an idea of any consistent patterns in the data and to help in decisions concerning the choice of a statistical framework within which to analyse the main data set, and, second, to use the limited amount of trial data to help teach the client (and other potential clients) how to look at and explore this type of data for himself.

As far as the first aim was concerned, the main properties of the data are fairly clearly shown by a plot of the survival curves obtained from Table 1 (see Figure 1). There is clearly a distinct 'shoulder' to the left of these curves, indicating that there appears to be a minimum recall time. That is, there is a brief period when a memory recall cannot be obtained. This period corresponds to a combination of the time to understand the stimulus word at presentation and the physical reaction time to press the bar once a recall has been achieved. The minimum response time varies from one subject to another and from one word type to another, but the variability is relatively minor compared with estimates of recall rates after the minimum recall times have been allowed for. This allowance can be

Fig. 1. Survival curves for pleasant (●) and unpleasant (▼) memories

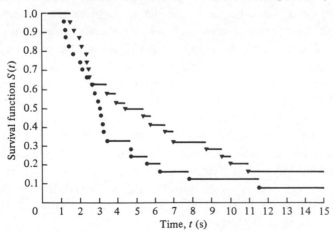

made for each memory type within each subject by deducting the minimum observed latency from all the observed latencies. Alternatively, a simpler rough-and-ready way involves subtracting 1 second for all observed latencies. The resulting times will be referred to as the transformed latencies.

A plot of the logarithm of the survival function against the transformed latencies derived from Table 1 gives two lines that are roughly straight, indicating that memories are being recalled at random with a constant recall rate; that is, the latencies follow the negative exponential distribution. The two types of memory recall within each subject can therefore be summarised by estimates of these rates. These estimates are obtained by dividing the total number of *successful* recalls by the sum of all of the appropriate transformed latencies (that is, including the times for censored observations although these did not lead to successful recalls).

We can now turn to the second aim of the preliminary analysis: to educate the client. At the onset of the study the client had approached me to discuss the analysis of a large set of data which, at that time, was still being collected (see Master, Lishman and Smith, 1983). In previous work (see Lloyd and Lishman, 1975 or Teasdale and Fogarty, 1979) the latencies had been summarised for each type of memory and for each subject separately by mean recall times. The mean recall times were calculated by dividing the total amount of time spent by the subject trying to achieve recalls of a particular type of memory by the total number of trials for that type of memory (that is, censored observations were treated as if the subjects actually had recalled a memory at the time of censoring). This appears to be quite a sensible approach to many statistically naive laymen and the main challenge facing the consultant statistician here was to persuade the client to use the censored observations in a different way from the uncensored ones (that is, to use the appropriate rate of recall derived from the transformed latencies rather than the mean recall time). The result of this persuasion has been that the client and his peers have adopted the more appropriate summary statistic, but in practice have always presented it in parallel with the defective one. Even though the collectors of this type of data admit that the crude mean recall time is defective, they are loath to abandon it *completely* in favour of a rate estimate. It is strange that even the median recall time, which would have been much more preferable to the arithmetic mean, does not appear to hold any attraction for psychiatrists.

But what difference does ignoring both the highly skewed distribution of the latencies and the fact that some of them are censored make? Not as much as I would have expected.

Consider, for example, the data in Table 1. If these latencies are transformed by replacing t by $t-1$ (assuming that the mean recall time is approximately 1), a log-linear exponential model can be fitted to the data using GLIM (Baker and Nelder, 1978; Aitkin and Clayton, 1980). The significance to the effect of changing from pleasant to unpleasant memories can be found by fitting either one survival curve or two survival curves to the data. G^2 for the difference in fit of the two models is 3.07 with 1 degree of freedom ($P < 0.08$). A simple t-test on the raw data yields a t-statistic of 1.53 with 46 degrees of freedom ($P < 0.06$ for a two-tailed test). Obviously, however, the discrepancy will increase as the number of censored observations increases.

4   The final product

The implication of the initial description of the various analyses carried out on Table 1 was that the larger body of data would be analysed using some form of proportional hazards model (Dunn and Master, 1983). However, a single-stage analysis of a large set of repeated measures of response latencies involving two or more within-subject contrasts (pleasant versus unpleasant; mild versus moderate effect; or mild versus strong effect), possible censoring and competing risks, together with the possibility of group influences (male versus female, for example), did not appear to be a very attractive proposition!

When the main data set was eventually ready to be analysed, the work was undertaken by a second consultant statistician, Alan Smith (see Master, Lishman and Smith, 1983). He chose to analyse the data in two stages. First, the latencies of each type were summarised by rate estimates as explained in Section 3 above. A development of the simple exponential model was made to allow for competing risks (David and Moeschberger, 1978). That is, each of the two types of memory (pleasant or unpleasant) were further classified into three sub-groups, depending on the rated intensity of recall. The rate for mildly pleasant memories, for example, can be estimated from the total number of mildly pleasant recalls and the total time spent trying to recall pleasant memories in general (moderately and very pleasant recalls are treated here as further examples of censoring).

To summarise, the first stage of the analysis yielded a set of six summary statistics for each subject (rates from mildly, moderately and very pleasant memories and for mildly, moderately and very unpleasant ones). The second stage of the analysis involved the use of simple significance tests to compare these rates both within and between subjects. The reader is referred to Master, Lishman and Smith (1983) for details. This two-stage approach is straightforward and relatively easy for the client to understand.

The second stage could have been made more complicated through the use of repeated-measures analysis of variance programs, but I doubt if this would have added much.

Following the publication of the paper by Master, Lishman and Smith (1983), I was again approached for help in the design and analysis of experiments involving the use of response latencies. It was assumed that the above two-stage procedure would be adopted as a standard method of approaching the analysis of any of the sets of data to be generated in future experiments. As a result of this it was decided that the microcomputer used to collect the latency data would also be programmed to carry out the initial estimation of the required recall rates (allowing for minimum response times, censoring and competing risks, if necessary). The output of the first stage of the analysis then formed the raw data for a second stage to be performed on a different (larger) computer using standard statistical packages such as SPSS.

The final product of the consultation with the statisticians was not the analysis of a particular set of latencies but a general strategy for their collection and subsequent analysis. The design of the experiments could be any one of the repeated-measures designs already familiar to experimental psychologists (see Winer, 1971). The analysis could involve the simple use of straightforward significance tests for specified within-subject or between-subject contrasts or the relatively sophisticated use of a repeated-measures analysis of variance/covariance program such as MANOVA in SPSSX (SPSS Inc., 1983).

5   Discussion

In the introduction to this chapter it was asked what the basic strategy should be in the analysis of repeated-measures designs. Should the main criterion, apart from validity, be one of simplicity? I have indicated in Sections 3 and 4 above how a set of data, which, when first encountered, would appear to defy a simple analysis, can be analysed in a fairly easy and straightforward way. There may be more technically perfect ways of approaching a complex set of survival data but I doubt whether the client or his peers would have much chance of following it.

In the context of the type of data set given as an example in this chapter, the client has to be able the understand and cope with several important problems.

    (i) The design of the experiment. This varies from one experiment to another, and of course, can be as simple or as complex as the client wishes it to be. My own view is that most of the experimental designs used in this area should be simpler than most clients would wish.

(ii) Censoring.

(iii) Competing risks (perhaps introduced as a generalisation of the idea of censoring, or vice versa).

(iv) Repeated measures (of two classes: repeated latencies of one type and parallel rate estimates for the different types of memory).

Once the *client* has understood the problems, the solutions of them can be quite straightforward. The solutions to the problems of repeated measurements will depend as much as anything on the tastes of the statistical consultant and on the statistical knowledge of the client. I, as a statistical consultant, prefer the simpler solutions, and I am sure that a statistically naive client does also. This does not mean, however, that the simple solutions can be provided without thought. If any corners are to be cut, if any simplifications to the analysis are to be made, then there is still a requirement for someone with sufficient experience and knowledge to know what simplifications are appropriate and what corners can be cut without threatening the validity of the resulting analysis. Simplicity does not imply that the statistical consultant becomes redundant, but that the consultant becomes more a source of advice and knowledge rather than another pair of hands to analyse the data for the client.

References

Aitkin, M. and Clayton, D. (1980) The fitting of exponential, Weibull and extreme value distributions to complex censored survival data using GLIM. *Applied Statistics*, **29**, 156–63.

Baker, R. S. and Nelder, J. A. (1978) *The GLIM System, Release 3; Generalised Linear Interactive Modelling*. Royal Statistical Society, London.

David, H. A. and Moeschberger, M. L. (1978) *The Theory of Competing Risks*. Griffin, London and High Wycombe.

Dunn, G. and Master, D. (1982) Latency models: the statistical analysis of response times. *Psychological Medicine*, **12**, 659–65.

Everitt, B. S. and Dunn, G. (1983) *Advanced Methods of Data Exploration and Modelling*. Heinemann Educational, London.

Lishman, W. A. (1974) The speed of recall of pleasant and unpleasant experiences. *Psychological Medicine*, **4**, 212–18.

Lloyd, G. G. and Lishman, W. A. (1975). Effect of depression on the speed of recall of pleasant and unpleasant experiencies. *Psychological Medicine*, **5**, 173–80.

Master, D., Lishman, W. A. and Smith, A. (1983) Speed of recall in relation to affective tone and intensity of experience. *Psychological Medicine*, **13**, 325–31.

SPSS Inc. (1983). *SPSSX User's Guide*. McGraw-Hill, New York.

Teasdale, J. D. and Fogarty, S. J. (1979) Differential effects of induced mood on retrieval of pleasant and unpleasant events from episodic memory. *Journal of Abnormal Psychology*, **88**, 248–57.

Winer, B. J. (1971) *Statistical Principles in Experimental Design* (2nd edition). McGraw-Hill, New York.

7

Acid rain and tree roots: an analysis of an experiment

J. N. R. JEFFERS

1 Introduction

The ideal form of statistical consultation starts with a discussion between the research worker and the statistician on the design of the whole research project, as well as the design of any component experiments or surveys. In this way, the statistician can ensure that all the data that are collected can be fitted together into a comprehensive model of the processes that underlie the phenomena being investigated. Only then does he help the research worker to identify the factors to be controlled or varied throughout the research, and the variables to be measured. The methods of mathematical analysis that will be used to test hypotheses, or to estimate parameters, then follow logically from the design of the whole project and its component experiments.

Unfortunately, very little consultation with statisticians follows this ideal pattern. Indeed, few research organisations have enough statisticians for more than a very small proportion of research projects to be planned with the active collaboration of the statisticians. What frequently happens, therefore, is that the statistician is consulted long after the data have been collected, and, by definition, long after the design of the investigation has been fixed and partly executed. It is also only very seldom that much thought will have been given to the integration of several (or many) experiments or surveys, so that, if the statistician is consulted at all, it will usually be about the analysis and interpretation of a single experiment or survey, without much regard for the whole investigation into which the individual research activities fit.

Increasingly, because of the ready availability of statistical packages on computers, research workers do not see the need to discuss their results with statisticians, and still less seek advice on the types of analysis which they should use. There are many reasons given for this apparent avoidance of consultation, ranging from the inaccessibility of statisticians, through the desire to reduce delays in the publication of results that may be imposed by statisticians who insist on going back to the origins of the research rather than answering a simple question, to a genuine lack of appreciation of the difficulties of the interpretation of data, especially when those data are derived from experimental or survey designs which are less than efficient, if not actually invalid. Nevertheless, it is an unfortunate fact that much of what reaches the statistician for advice only does so because the research worker has already encountered difficulties in the analysis or interpretation of the data. The statistician's main task is then to retrieve what he can from a situation which has often been made more difficult by a failure to impose effective standards on the design of research and on the collection of data. The scale of the problem, in the absence of direct advice from a collaborating statistician, can be seen by the number and content of the questions contained in simple checklists on experimental design and sampling (Jeffers, 1978, 1980).

Even where effective arrangements for the provision of statistical advice are made within a research organisation, the problem may be complicated by collaboration with other research organisations. Thus, one organisation may find itself obliged to work with designs which have been created by others, often without the benefit of advice. While it might be wiser on occasion to avoid getting into this situation, there are times when it is necessary to accept something which is less than ideal, and to make the best of the data which can be derived from the research. Such situations often provide a major challenge to the consultant statistician, as well as placing some constraints on what he would like to say about the design of the research generally.

The data described in this chapter were derived from an experiment designed by one organisation upon which was grafted an investigation by a second organisation. Only when the data had been collected was it realised by the second organisation that the interpretation of these data involved problems of analysis that had not been anticipated in the ways in which the data were collected. As is often the case in such situations, the problem is to extract the maximum information from the data that are now available rather than to advise on how the required information could be obtained efficiently and economically through an ideal design.

2   Background
         A good deal of concern has recently been focused on the problem of acid precipitation, and the effects of acid rain and sulphur deposition on plants (Jacobson, 1984). Much of the research on such effects has

concentrated on the above-ground parts of plants, but there is strong evidence that soils are also affected, as well as the roots of plants. Of particular importance is the possible effect of acid precipitation on long-lived organisms such as trees, and one theory is that acid precipitation has an adverse effect on the fine roots of such trees. Such an effect may be partly due to the direct damage of the roots themselves, but it may also be due to changes in the uptake of essential nutrients such as phosphates and nitrates.

Most fine roots of forest trees have an intimate relationship with symbiotic fungi known as mycorrhizas. Essentially, the role of these mycorrhizas is to enhance the nutrient uptake of the tree roots, possibly because of their ability to explore a greater volume of soil than the roots themselves and the associated root hairs. It was suspected that acidic pollutants might have an effect, not only on the survival of roots, but also on their mycorrhizal associates, possibly by selecting acid-tolerant species. Destruction of the root and mycorrhizal structures as a result of pollution would seriously affect the nutrient uptake of the trees, and research is therefore desirable on the extent of such effects.

The data included in the analysis described in this chapter were derived from an investigation of the effects of acid precipitation on the roots of seedlings of Scots pine. The seedlings were grown in four specially constructed lysimeters, each 0.8 m in diameter and 1.35 m deep, on an undisturbed humo-ferric podzol from the Lower Greensand series. All of the lysimeters received an artificial rain solution representing the composition of rain at Birkenes in south Norway, but for two of the lysimeters this artificial rain was acidified to pH 3.0 with sulphuric acid. All of the simulated rainfall applications were equivalent to an annual rainfall of 1500 mm, characteristic of south Norway.

The primary objectives of the investigation were:

  (i) to assess the effects of the acid solution on the biomass of the root system of the Scots pine seedlings;
  (ii) to assess the nature of the mycorrhizal symbionts in both the control and acid-treated lysimeters, and possibly to identify acid-tolerant species;
 (iii) to identify changes in the mycorrhizal structure due to acid treatment by examination of thin sections of fine root material;
 (iv) to compare the phosphate deficiency of surface roots in order to determine changes in root physiology due to the effects of the acid solution.

Three years after the initiation of the experiment, two soil cores, each with a diameter of 2.5 cm and an approximate depth of 35 cm, were taken from each of six colour-coded sectors of the lysimeters. The sampling positions within each sector were stratified, so that one core in each sector came from identically located points on the inner portion of the sector and the other core came from the outer portion of the sector. The litter layer from each sector was removed, and the core itself cut into 3 cm sections, corresponding to depth horizons, to a depth of 30 cm. The remaining fraction below 30 cm was kept as a separate sample. The litter and soil horizons were placed into individual pre-labelled self-seal polythene bags for transportation to the laboratory, where the cores were stored at 4 °C before subsequent processing.

The fine root length in each depth horizon and segment was estimated by a modified line intersect method with the aid of a Quantimet image analyser. The total number of root tips in each of seven mycorrhizal types was counted for each sample (Dighton, Skeffington and Brown, 1986). Because of the time involved in processing the samples, it was only possible to extract roots from one of the two cores taken from each segment. Alternate inner and outer cores were used, but this distinction between inner and outer samples is not used in the following analysis. A bioassay of the phosphorus uptake of the roots was also made, but only the root tip data are considered in this chapter.

3   Experimental design

The above description of the lysimeter experiment does not explicitly define any particular design as being appropriate to the analysis of the data derived from the experiment. In essence, there are two replications of each of the control and acid-treated lysimeters, and these could, perhaps, be regarded as following one of the possible six permutations of a 2×2 randomised block experiment (Figure 1). There is no evidence, however, that the assumed blocks have been used constructively to take up extraneous experimental errors. As an alternative, the allocation of the treatments could be regarded as completely randomised.

The division of the lysimeters into six sectors imposes a further restriction on the selection of the sample cores. Given the limited replication of the treatments to the lysimeters, it is tempting to use the differences between sectors within lysimeters as measures of the experimental error, but the systematic nature of the division into sectors poses problems for the analysis. The six cores from each lysimeter cannot be regarded as completely random samples from the lysimeters. The sectors were coded by colours indicating their position in relation to the centre of the lysimeter, and it is possible that the position of the sectors would reflect differences due to drainage or the application of the simulated precipitation.

The original intentions of the investigators of the effects of acid rain on the mycorrhizas was to analyse the root tip data by a four-factor analysis of variance, the four factors being treatment (control versus acid), sectors, depths, and mycorrhizal type. However, it quickly became apparent that the sectors, depths and mycorrhizal types could not be regarded as experimental factors in such an analysis. Rather, the recorded number of root tips at each 3 cm depth horizon belonging to each of the mycorrhizal types should be regarded as the variables of a 7×11 multivariate set. Any efficient analysis of the data, therefore, should acknowledge the existence of the whole set of variables, as opposed to simply analysing each variable separately.

4   Exploratory analysis

The first step in any statistical analysis today is to make the data machine readable to allow the appropriate analysis to be done quickly and conveniently on a computer. The data in Table 1 were therefore stored on

Fig. 1. Lysimeter layout (sectors coded in colours: G green; B blue, V violet; Y yellow; R red; O orange)

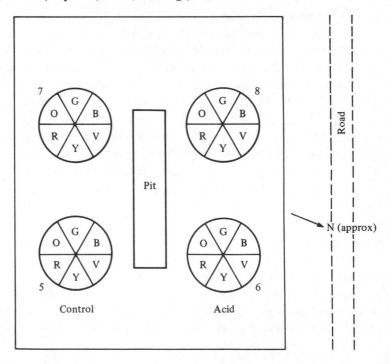

Table 1. *Effects of acid rain on Scots pine seedlings, mycorrhizal types*

	Type	Depths (cm)										
		0–3	3–6	6–9	9–12	12–15	15–18	18–21	21–24	24–27	27–30	30–33
Lysimeter 5, control												
	A	219	30	26	86	25	25	0	0	0	12	0
	B	171	23	29	18	22	21	0	0	0	0	0
	C	74	83	22	16	124	58	0	0	0	0	0
Y	D	0	0	0	2	19	0	0	0	0	0	0
	E	11	11	16	0	0	0	0	0	0	1	0
	F	0	0	0	8	24	116	40	0	0	0	0
	G	33	5	3	1	1	2	0	0	1	0	0
	A	539	94	144	62	77	0	15	18	5	22	8
	B	0	0	0	3	0	0	0	0	0	0	0
	C	80	38	25	38	53	0	27	11	5	0	2
V	D	0	0	0	0	0	0	0	0	0	0	0
	E	87	86	73	59	0	0	0	0	0	0	3
	F	358	0	0	188	0	0	0	0	0	0	0
	G	4	2	5	4	2	0	0	0	3	0	0
	A	548	106	57	12	16	0	0	0	0	0	0
	B	833	452	95	0	108	0	0	0	0	0	0
	C	133	172	24	2	4	0	0	0	0	0	0
O	D	0	0	43	0	0	0	0	0	0	0	0
	E	53	8	1	0	5	0	0	0	0	0	0
	F	4	20	0	0	0	0	0	0	0	0	0
	G	4	3	3	0	1	0	0	0	0	0	0
	A	452	55	146	51	9	35	40	28	14	20	5
	B	75	0	0	0	0	0	0	0	0	0	0
	C	138	119	88	22	13	13	5	10	0	13	1
B	D	0	27	0	0	0	0	0	0	0	0	0
	E	13	7	0	0	0	0	0	0	0	0	0
	F	298	12	148	0	0	0	0	8	0	0	0
	G	10	4	1	0	0	1	1	1	0	0	0
	A	330	110	88	33	17	26	8	23	0	21	34
	B	28	0	0	0	0	0	0	10	0	42	4
	C	155	80	38	30	15	8	0	5	0	0	4
G	D	125	0	0	0	0	0	0	0	0	0	0
	E	20	13	0	0	0	0	0	0	0	0	0
	F	358	26	0	0	0	0	0	0	0	0	0
	G	3	2	1	2	4	0	0	1	0	0	0
	A	287	107	12	20	16	0	0	0	0	31	10
	B	426	465	411	31	2	0	0	0	0	24	2
	C	36	19	7	18	12	0	0	3	0	5	5
R	D	0	0	11	0	0	0	0	0	0	3	0
	E	23	2	1	1	0	0	0	0	0	2	0
	F	57	80	0	0	0	0	0	0	0	0	0
	G	0	1	1	0	0	0	0	3	0	0	0

Table 1. (*cont.*)

| | Type | \multicolumn{11}{c}{Depth (cm)} |
		0–3	3–6	6–9	9–12	12–15	15–18	18–21	21–24	24–27	27–30	30–33
Lysimeter 7, control												
Y	A	493	34	52	0	0	0	0	0	0	4	33
	B	60	0	0	0	0	0	0	0	0	0	0
	C	298	12	21	1	0	0	0	0	0	1	18
	D	49	0	0	0	0	0	0	0	0	0	0
	E	22	0	0	0	0	0	0	0	0	0	0
	F	0	0	0	0	0	0	0	0	0	0	0
	G	4	0	1	0	0	0	0	0	0	0	0
V	A	306	36	161	111	29	4	1	19	72	5	1
	B	459	0	8	0	0	0	0	0	0	0	0
	C	85	114	61	41	8	1	9	27	14	7	12
	D	49	0	0	0	0	0	0	0	0	0	0
	E	5	20	13	5	1	0	0	0	0	3	0
	F	0	0	0	0	0	0	0	0	20	0	0
	G	0	2	1	1	0	0	0	1	0	0	0
O	A	546	73	87	6	2	0	0	0	0	14	0
	B	315	6	7	0	0	0	0	0	0	0	0
	C	8	67	19	8	15	3	0	0	0	1	0
	D	0	0	0	0	0	0	0	0	0	0	0
	E	155	3	0	0	0	0	0	0	0	0	0
	F	0	0	0	0	0	0	0	0	0	0	0
	G	3	0	0	0	1	1	0	0	0	0	0
B	A	215	92	96	45	0	0	0	0	19	79	9
	B	620	19	145	0	0	0	0	0	0	0	2
	C	23	15	23	31	4	0	0	0	0	3	4
	D	0	0	11	8	0	0	0	0	0	0	0
	E	58	13	4	0	0	0	0	0	20	65	0
	F	0	0	0	0	0	0	0	0	0	4	1
	G	3	1	0	2	0	2	0	0	0	2	0
G	A	383	146	138	88	40	0	3	103	31	20	12
	B	247	8	21	0	0	0	0	0	23	0	0
	C	112	156	63	26	28	0	20	27	15	20	2
	D	0	0	0	0	0	0	0	0	0	0	0
	E	191	36	0	0	0	0	0	1	16	4	1
	F	0	0	2	0	0	0	0	0	0	0	0
	G	5	3	3	0	0	0	0	0	0	0	1
R	A	281	68	3	24	0	0	0	0	34	0	0
	B	580	70	0	2	0	0	0	8	0	0	0
	C	59	81	6	79	0	0	0	0	13	4	0
	D	0	0	0	0	0	0	0	0	0	0	0
	E	168	48	1	15	0	0	0	0	2	0	0
	F	0	0	0	0	0	0	0	0	12	0	0
	G	6	0	0	0	0	0	0	0	1	0	0

Table 1. (*cont.*)

		Depth (cm)										
	Type	0–3	3–6	6–9	9–12	12–15	15–18	18–21	21–24	24–27	27–30	30–33
Lysimeter 6, acid												
Y	A	247	95	86	3	4	0	8	46	15	0	0
	B	19	9	0	0	0	0	0	0	0	0	0
	C	15	9	28	1	5	9	0	15	5	0	0
	D	0	0	0	0	0	0	0	0	0	0	0
	E	11	9	1	0	0	0	0	0	0	0	0
	F	0	0	3	0	0	0	0	0	0	0	0
	G	9	1	3	0	0	0	0	1	0	0	0
V	A	369	63	11	48	76	7	7	0	20	12	11
	B	0	0	15	1	0	0	0	0	0	0	5
	C	67	20	1	13	24	2	0	0	3	18	0
	D	0	0	0	0	0	0	0	0	0	0	0
	E	17	11	0	5	2	0	0	0	0	0	0
	F	41	7	0	0	0	0	0	0	0	0	0
	G	3	4	1	2	1	1	0	1	0	0	0
O	A	19	7	20	0	0	5	5	9	4	0	0
	B	0	0	3	0	0	0	0	0	2	0	0
	C	15	11	5	0	0	36	32	56	12	1	0
	D	0	0	0	0	0	0	0	0	0	0	0
	E	6	4	0	0	0	0	25	33	23	1	0
	F	0	0	0	0	0	0	0	0	0	0	0
	G	1	2	0	0	0	1	0	0	1	0	0
B	A	318	54	38	49	82	65	31	12	22	20	16
	B	40	0	4	0	0	0	0	0	0	0	0
	C	105	38	40	68	7	2	0	4	2	0	2
	D	0	0	0	0	0	0	0	0	0	0	0
	E	98	88	8	10	6	3	3	4	0	0	0
	F	0	6	0	0	0	0	0	0	0	0	0
	G	1	1	4	0	0	0	0	0	0	0	0
G	A	253	105	163	42	2	11	12	0	30	11	23
	B	6	0	0	0	0	0	0	0	0	0	0
	C	61	22	1	17	10	12	0	0	6	0	12
	D	16	0	22	0	0	0	0	0	0	0	0
	E	51	95	71	65	5	0	9	0	27	7	0
	F	7	0	0	0	0	0	0	0	0	0	0
	G	4	0	1	0	0	0	3	0	0	1	0
R	A	52	2	3	8	0	0	4	0	9	3	0
	B	5	0	0	0	0	0	0	0	0	0	0
	C	46	36	6	66	4	0	4	0	0	0	0
	D	0	129	0	6	0	0	0	0	0	0	0
	E	3	0	6	0	0	0	0	0	0	0	0
	F	11	0	0	0	0	0	0	0	0	0	0
	G	3	10	1	1	0	0	0	1	0	0	0

Table 1. (*cont.*)

	Type	0–3	3–6	6–9	9–12	12–15	15–18	18–21	21–24	24–27	27–30	30–33
						Depth (cm)						

Lysimeter 8, acid

	Type	0–3	3–6	6–9	9–12	12–15	15–18	18–21	21–24	24–27	27–30	30–33
	A	344	42	269	44	0	0	16	39	37	13	0
	B	3	0	5	12	0	0	0	0	6	2	0
	C	17	8	16	7	5	23	10	2	11	7	0
Y	D	4	53	0	0	0	0	0	0	13	0	0
	E	46	13	35	0	0	0	0	5	6	0	0
	F	0	0	0	0	0	0	0	0	0	0	0
	G	5	1	1	0	1	0	0	0	4	2	0
	A	440	4	142	58	0	0	0	0	0	14	0
	B	10	5	0	0	0	0	0	0	0	0	0
	C	163	6	20	1	3	0	0	0	0	1	0
V	D	11	0	0	0	0	0	0	0	0	0	0
	E	77	1	7	4	0	0	0	0	0	0	0
	F	0	0	0	0	0	0	0	0	0	0	0
	G	6	0	0	0	0	0	0	0	0	0	0
	A	98	68	20	10	79	43	26	41	28	0	6
	B	3	0	0	0	0	0	0	0	0	0	0
	C	493	532	55	12	11	2	3	15	5	0	2
O	D	10	0	0	0	0	0	0	0	0	0	0
	E	37	32	19	13	8	6	4	4	0	1	0
	F	0	0	0	0	0	0	0	0	0	0	0
	G	8	3	3	0	1	0	0	1	2	0	0
	A	50	46	359	25	7	15	4	0	16	36	1
	B	0	5	0	0	0	0	0	0	0	0	0
	C	44	72	108	15	6	3	5	0	12	0	0
B	D	34	16	0	0	0	0	0	0	0	0	0
	E	37	11	8	0	1	0	0	0	0	0	0
	F	0	0	12	0	0	0	0	0	0	0	0
	G	0	0	0	0	0	0	0	0	0	0	0
	A	130	72	60	0	0	0	0	0	21	2	3
	B	2	0	0	0	0	0	0	0	0	0	0
	C	107	8	36	0	0	0	0	0	0	0	6
G	D	93	81	0	0	0	0	0	0	0	0	0
	E	41	12	1	0	0	0	0	0	0	0	0
	F	0	0	0	0	0	0	0	0	0	0	0
	G	6	3	0	0	0	0	0	0	0	0	0
	A	215	108	44	19	0	0	31	69	6	14	7
	B	0	0	2	0	0	0	0	0	0	0	0
	C	139	32	28	8	0	0	59	75	3	1	0
R	D	12	0	0	0	0	0	0	0	0	0	0
	E	75	9	0	0	0	0	0	0	0	0	0
	F	0	0	0	0	0	0	0	0	0	0	0
	G	4	4	0	1	1	0	0	0	2	0	0

a floppy disk after being input and checked on a microprocessor. Before proceeding to any formal analysis, however, it is the author's strong preference to do some exploratory data analysis to see what the data themselves say about the nature of the variation which has been generated by the experiment.

The simplest way of summarising the 7×11 matrix of observations for each segment is to count the total number of root tips in the core, and these total numbers are given in Table 2 and Figure 2. Although these totals are not strictly random samples from each lysimeter, Figure 3 shows the range of variation in the totals of a box-plot diagram (Tukey, 1977). Clearly, the total range of variation does not vary markedly from lysimeter to

Table 2. *Total numbers of root tips in sectors of lysimeters*

	Control		Acid-treated	
Sector	5	7	6	8
Yellow	1378	1103	657	1137
Violet	2140	1712	889	973
Orange	2707	1340	339	1694
Blue	1893	1638	1251	950
Green	1664	1995	1183	684
Red	2134	1565	419	968

Fig. 2. Total number of root tips

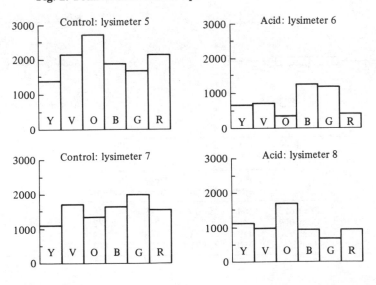

lysimeter, but the mid-spread for the acid-treated lysimeter 6 is larger than for the other three lysimeters, and acid-treated lysimeter 8 has a very small mid-spread which effectively makes the two extreme values outliers to the remaining four values. The medians of the two acid-treated lysimeters are less than those of the two control lysimeters, suggesting a reduction in the total number of root tips to about half of those found in untreated soils.

An alternative presentation of the total numbers of root tips is given in Table 3 as a median polish (McNeil, 1977) of the two-way relationship between sectors and lysimeters. Again, there are marked differences between the components for the acid-treated and control lysimeters, but the analysis also suggests that there are small, but consistent, effects of the sectors. The residuals from the two-way median polish are plotted in Figure 4, but do not suggest any particular pattern, even after trying various alternative configurations of the sectors to reflect possible drainage patterns within the lysimeters.

Fig. 3. Range of variation in total number of root tips: box-plot diagrams

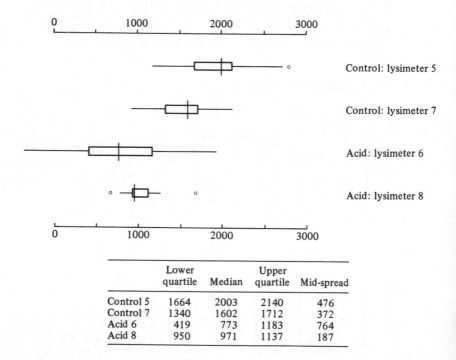

	Lower quartile	Median	Upper quartile	Mid-spread
Control 5	1664	2003	2140	476
Control 7	1340	1602	1712	372
Acid 6	419	773	1183	764
Acid 8	950	971	1137	187

The average numbers of root tips in each 3 cm depth horizon for each lysimeter are given in Table 4 and plotted in Figure 5. The numbers of root tips decline rapidly with depth, from just under 1000 at a depth of 3 cm in the control lysimeters and less than 500 in the acid-treated lysimeters to less than 50 at the lower depths in both control and acid-treated soils. The principal effect of the acid treatment, therefore, is to reduce the numbers of root tips in the top 3 cm of the soil.

Finally, the numbers of root tips with different mycorrhizal associations in each lysimeter are given in Table 5. The principal effect of the acid treatment appears to have been a large reduction in the numbers of root tips associated with types B and F, and a rather smaller reduction of those with type A. It is notable, however, that only one of the control lysimeters

Table 3. *Two-way median polish of total numbers of root tips*

	Control		Acid-treated		
Sector	5	7	6	8	Typical value
Yellow	−224	−127	101	549	−380
Violet	104	48	−100	−48	54
Orange	497	−498	−825	448	228
Blue	−104	23	310	−23	5
Green	−313	390	252	−279	−5
Red	174	−23	−494	23	−22
Typical values	693	321	−353	−321	1289

Fig. 4. Residuals for median polish

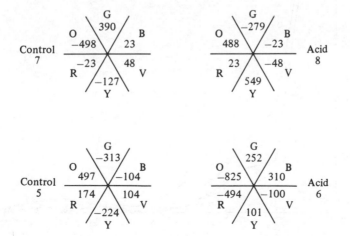

had any substantial numbers of root tips associated with type F. Table 5 does not show the effect of depth on the changes in numbers of root tips, but, as most of the effects of the acid treatment occurred in the top 3 cm of the soil, it can be assumed that most of the effects on the root tips with different mycorrhizal associations occurred near the surface.

Table 4. *Average numbers of root tips in each 3 cm horizon of cores*

Depth (cm)	Control		Acid-treated	
	5	7	6	8
0–3	998	968	320	461
3–6	377	187	140	206
6–9	255	158	91	208
9–12	121	82	68	38
12–15	110	21	38	21
15–18	38	2	26	15
18–21	16	6	24	26
21–24	20	31	30	42
24–27	5	49	30	30
27–30	33	39	12	16
30–33	13	16	12	4

Fig. 5. Number of root tips in 3 cm horizons

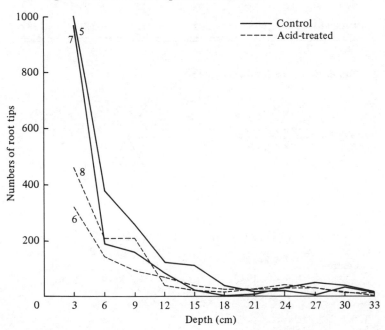

The exploratory analysis, therefore, suggests a marked reduction in the total number of root tips of the Scots pine seedlings after three years as a result of the acid treatment. Most of this reduction took place in the top 3 cm of the soil, and is related to three of the possible seven mycorrhizal associations. What we now require is a more thorough analysis of the data, providing precise estimates of the effects. Ideally, we would also like a reasonably simple model of the experimental system which would perhaps be used to make predictions against which to plan future research.

5   Analysis

The formal analysis of the data of Table 2, which will now be addressed, illustrates quite well the problems posed by the design of the experiment. It will be worthwhile, therefore, to spend a little time examining some alternative analyses in more detail.

If we assume that the control and acid treatments were allocated to the four lysimeters at random, and that the cores from the six sectors in each lysimeter also represent random samples of the soil in the lysimeters, the

Table 5. *Average numbers of root tips for mycorrhizal types*

Mycorrhizal type	Control		Acid-treated	
	5	6	7	8
A	696	670	440	542
B	549	433	18	9
C	309	278	162	366
D	36	20	29	55
E	83	145	118	84
F	291	7	13	2
G	19	7	11	10

Table 6. *Analysis of variance of total numbers of root tips*

Source of variation	Degrees of freedom	Sum of squares	Mean square	F-ratio
Controls versus acid	1	4263051.04	4263051.04 ⎫	11.00
Between lysimeters	2	775002.75	387501.38 ⎭	
Within lysimeter 5	5	1049401.33	209880.27	
Within lysimeter 7	5	475678.83	95135.77	
Within lysimeter 6	5	735501.33	147100.27	
Within lysimeter 8	5	577041.33	115408.27	

appropriate analysis of variance of the untransformed total numbers of root tips per core is as in Table 6. One possible test of the mean difference in numbers of root tips per core between control and treated soils is given by the F-ratio

$$F(1, 2) = 4263051.04/387501.38 = 11.00$$

While this value of the F-ratio looks encouragingly high, the very small number of degrees of freedom for the denominator of this ratio effectively precludes the mean difference being shown as statistically significant. Expressing the same result in another way, the mean difference between the numbers of root tips per core in control and acid-treated soils is estimated by

$$1771.6 - 928.7 = 842.9$$

with a standard error estimated by $\sqrt{(387501.38\sqrt{12})} = 179.7$, but the size of the difference necessary for it to be shown as significant, even at the 5% level of probability, is equivalent to

$$179.7 \times \sqrt{2} \times t(2 \text{ d.f.}) = 179.7 \times \sqrt{2} \times 4.303 = 1093.5$$

The lack of replication of treated and untreated lysimeters makes it impossible to ascribe significance to the effects of treatments based on differences between lysimeters treated alike.

An alternative measure of variability, increasing the number of degrees of freedom for a test of significance, or an estimate of the mean difference, is the 'within lysimeter' variance derived from the individual cores. However, before proceeding, it is first necessary to test the homogeneity of the 'within lysimeter' variances. For the control lysimeters, the F-ratio

$$F(5, 5) = 209880.27/95135.77 = 2.21$$

does not exceed the tabulated value of F for the 5% level of probability of 7.15. The two variances may therefore reasonably be pooled to give an estimated common variance of 152508.01 for the 'within lysimeter' variability of the control lysimeters. Similarly, for the acid-treated lysimeters, the F-ratio

$$F(5, 5) = 735501.33/577041.33 = 1.27$$

is again not significant at the 5% probability level and the two variances may be pooled to give an estimated 131254.27 for the common 'within lysimeter' variance of the acid-treated lysimeters. Finally, the F-ratio for the comparison between the control and acid-treated variances is given by

$$F(10, 10) = 152508.01/131254.27 = 1.6$$

and suggests that all of the variances can be pooled to give a common 'within lysimeter' variance of 141881.14. Alternatively, the heterogeneity of all four variances can be tested simultaneously by Bartlett's test

(Snedecor, 1946) giving an adjusted chi-square of 0.82 with 3 degrees of freedom, again not significant even at the 5% level of probability.

This pooled 'within lysimeter' or 'between cores' variance enables two further F-ratio tests to be performed. First, the F-ratio of the comparison of control and acid-treated lysimeters is given by

$$F(1, 20) = 4263051.04/141881.15 = 30.04$$

which far exceeds the tabulated value of F for the 1% probability level. Moreover, the F-ratio for the differences between lysimeters treated alike is given by

$$F(2, 20) = 387501.38/141881.14 = 2.73$$

which does not exceed the tabulated value of F for the 5% probability level. If, therefore, the 'within lysimeter' variances can be regarded as valid estimates of the 'between core' variances, the difference between the numbers of root tips per core in the control and acid-treated soils is statistically significant, and there are no significant differences between the numbers of root tips per core in the soils of lysimeters treated alike. The difference in numbers of root tips per core between the control and acid-treated soils is estimated as

$$1771.6 - 928.7 = 842.9$$

with a standard error

$$\sqrt{(2 \times 141881.14/12)} = 153.8$$

The reduction in the number of root tips per core resulting from treatment with dilute sulphuric acid is approximately 48%. As the volume of each core was 161.7 cm, these total numbers can be readily converted to numbers of root tips per cubic centimetre, and Table 7 summarises the results of the analysis so far.

The above analysis assumes that the 'within lysimeter' samples were taken at random within each lysimeter. In fact, the randomisation of the samples was restricted, two cores being taken from each of six sectors coded by colours in Figure 1. For the data of Table 1, only one of these cores was examined in detail, and the data therefore represent the numbers

Table 7. *Summary of numbers of root tips per core and per cubic centimetre*

Treatment	Total no. of root tips per core	Total no. of root tips per cm³
Control	1772 ± 105	11.0 ± 0.7
Acid	929 ± 109	5.7 ± 0.7
Difference	843 ± 154	5.2 ± 1.0

of root tips found in each of one core taken at random from each of the six sectors. As it is quite likely that there would be some effect of the orientation of the sectors on the distribution of the numbers of root tips, it is appropriate to examine the numbers of root tips in each sector, and Table 8 gives an analysis of variance identifying the sectors, and the interaction between the effects of the sectors and the comparison between the control and treated soils. Because the sectors represent a systematic division of the surface area of the lysimeters, it is necessary to calculate two 'error' terms (Cochran and Cox, 1950). One of these 'errors' applies to the comparison of sectors and the treatment. The appropriate F-ratio tests

$$F(5, 5) = 102238.85/17853.30 = 5.73$$

and

$$F(5, 5) = 86753.45/360696.97 = 0.24$$

indicate a significant ($P = 0.05$) difference between the numbers of root tips per core in the sectors, but no significant interaction between the effects of the sectors and the acid treatment. Table 9 and Figure 6 show the numbers of root tips per core for each sector, and indicate that the numbers of root rips were significantly less in the yellow sectors of the lysimeters than in some of the other sectors. No obvious reason has been advanced for this difference, and, fortunately, it has not interacted with the effect of the acid treatment.

6   Effect of depth

The preliminary analysis of the data summarised in Table 4 and Figure 5 suggested a marked reduction in the number of root tips with depth in the soil, but did not seek to characterise the relationship between the numbers of root tips and depth. Inspection of the curves representing

Table 8. *Analysis of variance testing differences between sectors*

Source of variation	Degrees of freedom	Sum of squares	Mean square	F-ratio
Control versus acid	1	4263051.04	4263051.04	
Between lysimeters	2	775002.75	387501.38	
Sectors	5	511194.25	102238.85 ⎫	5.73
Error 1	5	89176.50	17853.30 ⎭	
Treatments versus sectors	5	433767.25	86753.45 ⎫	
				0.24
Error 2	5	1803484.90	360696.97 ⎭	
Total	23	7875676.61		

Table 9. *Average numbers of root tips per core in colour-coded sectors*

Sector	No. of root tips per core
Yellow	1069
Red	1272
Green	1382
Violet	1429
Blue	1431
Orange	1520
Standard error	66.8

Fig. 6. Diagrammatic representation of numbers of root tips in each sector

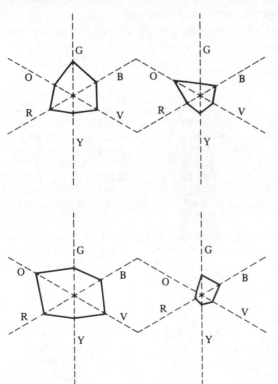

the average number of root tips per core in each 3 cm horizon suggests a range of possible relationships, including exponential, logarithmic and hyperbolic regressions, for example

Exponential $\quad y = a + b \ln x$

Logarithmic $\quad \ln y = a + bx$

Hyperbolic $\quad \ln y = a + b \ln x$

Table 10 shows the proportion of the variability in the numbers of root tips per horizon accounted for by six possible regression relationships. Clearly, none of these regressions is 'best' for all of the lysimeters, but overall the regression of the logarithm of the number of root tips on the logarithm of depth accounts for a consistently high proportion of the variability.

Table 11, therefore, gives the calculated values of the regression constant and coefficient for the relationship between the logarithm of the number of root tips on the logarithm of depth for each lysimeter, together

Table 10. *Variability accounted for by six possible regressions of numbers of root tips on depth of soil horizon*

| | Proportion of total variability | | | |
| | Control | | Acid | |
Regression	5	7	6	8
$\ln y$ on x	0.820	0.343	0.863	0.737
y on $\ln x$	0.836	0.672	0.857	0.833
$\ln y$ on $\ln x$	0.861	0.532	0.944	0.796
$\sqrt{y}$ on x	0.751	0.479	0.756	0.676
y on $\sqrt{x}$	0.697	0.517	0.724	0.707
$\sqrt{y}$ on $\sqrt{x}$	0.861	0.615	0.863	0.781

Table 11. *Parameters of regressions of log numbers of root tips on log depth for each lysimeter, and for control and acid-treated lysimeters*

Lysimeter	Proportion of variability	Regression constant	Regression coefficient
5	0.861	9.54119	−2.03032
7	0.532	8.24055	−1.68775
6	0.944	7.31835	−1.32647
8	0.796	8.14692	−1.65012
Control	0.872	8.82215	−1.76511
Acid	0.892	7.75491	−1.48371

with the correlation coefficients. Except for lysimeter 7, for which the regression accounts for only about 53% of the variation, the regressions generally account for at least 80% of the variation in the number of root tips. However, as can be seen from the table of residuals in Table 12, the fitted curves are not entirely satisfactory because the pattern of residuals shows systematic departures from the predicted values. The same relationship is fitted to the average numbers of root tips in each horizon for the control and acid-treated lysimeters in Table 13. Again, while the general trend of the fitted curves is approximately right, there are systematic patterns in the residuals from the fitted values.

Table 12. *Residuals from fitted regressions of log numbers of root tips on log depth for individual lysimeters and averages of control and acid-treated lysimeters*

Depth (cm)	Residual (actual − fitted)					
	5	7	6	8	Control	Acid
3	498.7	−374.3	31.3	102.8	−7.2	66.8
6	−10.7	−2.9	0.3	−26.7	4.9	−9.5
9	−93.7	−64.9	−9.1	−116.4	−65.9	−60.0
12	−30.8	−25.0	−11.7	19.0	−16.9	5.6
15	−53.2	17.9	3.5	18.8	−8.8	12.6
18	1.2	27.0	6.7	14.0	21.3	11.5
21	12.8	16.8	2.7	−3.6	20.7	0.4
24	1.8	−13.2	−7.7	−23.6	−0.7	−15.2
27	12.6	−34.1	−11.1	−15.3	−6.5	−12.7
30	−19.7	−27.2	4.2	−2.9	−19.3	1.1
33	−1.5	−5.6	3.1	6.6	−0.3	5.2

Table 13. *Total number of root tips in 3 cm horizons*

Depth (cm)	Control	Acid
0–3	982.75	390.25
3–6	282.08	172.92
6–9	206.17	149.58
9–12	101.33	52.83
12–15	65.75	29.42
15–18	20.00	20.50
18–21	10.75	25.08
21–24	25.58	36.08
24–27	26.67	30.25
27–30	36.00	13.92
30–33	14.50	7.83

While it is possible to fit the actual values by polynomial curves of up to the fifth degree, these curves give very little insight into the underlying ecology of the distribution of root tips with depth. The principal question is whether the slight increases in numbers of root tips at the lower depths are 'real'. A variety of possible forms of analysis might be suggested, including the fitting of series of exponential curves representing mycorrhizas associated with particular strata, but such exercises are essentially peripheral to the main purpose of interpreting the results of the experiment.

Table 14, therefore, gives the average numbers of root tips per cm³ in each horizon, together with the appropriate standard errors (SE), derived from a series of analyses of variance similar to that given in Table 8 for total numbers of root tips. Expressed as numbers per cubic centimetre, the density of root tips declined rapidly from 67 at the surface to 4.5 at a depth of 12–15 cm. The only statistically significant differences in numbers of root tips, however, was in the surface 0–3 cm. Below a depth of 15 cm, there were only one or two root tips per cubic centimetre in both sets of lysimeters.

7   Effect of acid treatment on numbers of root tips with different mycorrhizal associations and at different depths

The numbers of root tips with different mycorrhizal associations and at varying depths for both control and acid-treated lysimeters are given in Tables 15–18 and Figure 7. The effect of the acid was to reduce the numbers of root tips with mycorrhizal associations A, B and F, but this effect was almost entirely confined to the top 6 cm of the soil. Below 6 cm, there were no significant effects of the acid treatment.

Table 14. *Number of root tips per cubic centimetre in 3 cm horizons*

Depth (cm)	Control	Acid	SE
0–3	66.9	26.5	4.48
3–6	19.2	11.6	3.58
6–9	14.0	10.2	2.51
9–12	6.9	3.6	1.47
12–15	4.5	2.0	1.26
15–18	1.4	1.4	0.65
18–21	0.7	1.7	0.46
21–24	1.7	2.5	0.88
24–27	1.8	2.1	0.56
27–30	2.4	0.9	0.64
30–33	1.0	0.5	0.28

As for the total numbers of root tips, the consistently best fit for the average number of root tips with any one mycorrhizal association was given by a power law regression of the form

$$y = ax^b$$

The parameters of the fitted regressions for control and acid treatments, and for each of the mycorrhizal associations are given in Table 19.

Table 15. *Average numbers of root tips by depth and mycorrhizal types: control*

| Depth (cm) | Mycorrhizal type | | | | | | |
	A	B	C	D	E	F	G
0–3	383	318	100	19	67	90	6
3–6	79	87	80	2	21	12	2
6–9	84	60	33	5	9	13	2
9–12	45	5	26	1	7	18	1
12–15	19	11	23	2	1	10	1
15–18	7	2	7	0	0	3	1
18–21	6	0	5	0	0	0	0
21–24	16	2	7	0	0	0	1
24–27	15	2	4	0	3	3	0
27–30	19	6	5	0	6	0	0
30–33	9	1	4	0	0	0	0

Fig. 7. Effects of acid treatment on proportional numbers of root rips with different mycorrhizal types

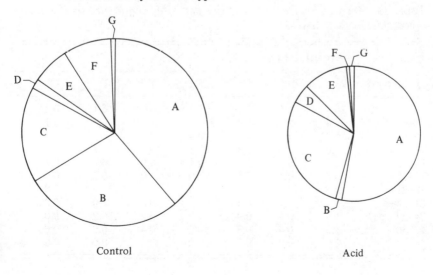

Control   Acid

8   Conclusions

Fortunately, the results of the experiment were fairly clear-cut. Simulation of acid rain by a weak solution of sulphuric acid reduced the number of root tips in Scots pine seedlings from 1770 to 930 in a complete core with a diameter of 2.5 cm and a depth of 35 cm. However, the effect of the acid treatment was confined to the surface layers, reducing the numbers of root tips from $77/cm^3$ in untreated soils to $27/cm^3$ in the top 3 cm in the acid-treated soils and from $20/cm^3$ to $12/cm^3$ between 3 cm

Table 16. *Average numbers of root tips by depth and mycorrhizal types*: *acid*

Depth (cm)	Mycorrhizal type						
	A	B	C	D	E	F	G
0–3	211	7	106	15	42	5	4
3–6	56	2	66	23	23	1	2
6–9	101	2	29	2	13	1	1
9–12	26	1	17	1	8	0	0
12–15	21	0	6	0	2	0	1
15–18	12	0	7	0	1	0	0
18–21	12	0	9	0	3	0	0
21–24	18	0	14	0	4	0	0
24–27	18	1	5	1	5	0	1
27–30	10	0	2	0	1	0	0
30–33	6	0	2	0	0	0	0

Table 17. *Average numbers of root tips per cubic centimetre by depth and mycorrhizal type*: *control*

Depth (cm)	Mycorrhizal type						
	A	B	C	D	E	F	G
0–3	26.1	21.6	6.8	1.3	4.6	6.1	0.4
3–6	5.4	5.9	5.4	0.2	1.4	0.8	0.1
6–9	5.7	4.1	2.3	0.4	0.6	0.9	0.1
9–12	3.0	0.3	1.8	0.1	0.5	1.2	0.1
12–15	1.3	0.7	1.6	0.1	0.0	0.7	0.1
15–18	0.5	0.1	0.5	0	0	0.2	0.0
18–21	0.4	0	0.3	0	0	0	0.0
21–24	1.1	0.1	0.5	0	0.0	0	0.0
24–27	1.0	0.1	0.3	0	0.2	0.2	0.0
27–30	1.3	0.4	0.3	0.0	0.4	0.0	0.0
30–33	0.6	0.0	0.3	0	0.0	0.0	0.0

and 6 cm. Below 15 cm, the acid treatment had no significant effect on numbers of root tips of the Scots pine seedlings. Moreover, the effect of the weak sulphuric acid solution was largely confined to root tips with mycorrhizal associations of types A, B and F.

In a very real sense, the experimenters were lucky. If there had been significant interactions between the effects of the acid treatment, depths and mycorrhizal types, it would have been very difficult to interpret those interactions, given the limited replication of the control and acid treatment, and the systematic sampling of the cores within the lysimeters. If, in

Table 18. *Average numbers of root tips per cubic centimetre by depth and mycorrhizal types: acid*

Depths (cm)	Mycorrhizal type						
	A	B	C	D	E	F	G
0–3	14.4	0.5	7.2	1.0	2.8	0.3	0.3
3–6	3.8	0.1	4.5	1.6	1.6	0.1	0.2
6–9	6.9	0.2	2.0	0.1	0.9	0.1	0.1
9–12	1.7	0.1	1.2	0.0	0.5	0	0.0
12–15	1.4	0	0.4	0	0.1	0	0.0
15–18	0.8	0	0.5	0	0.1	0	0.0
18–21	0.8	0	0.6	0	0.2	0	0.0
21–24	1.2	0	0.9	0	0.3	0	0.0
24–27	1.2	0.0	0.3	0.1	0.3	0	0.1
27–30	0.7	0.0	0.2	0	0.1	0	0.0
30–33	0.4	0.0	0.1	0	0	0	0

Table 19. *Parameter of power law equations for the regressions of numbers of root tips per horizon on lower depth of horizon*

Mycor-rhizal type	Control			Acid		
	Regression constant	Regression coefficient	Proportion of variability	Regression constant	Regression coefficient	Proportion of variability
A	1536	−1.515	0.801	876	−1.832	0.860
B	5123	−2.421	0.854	17	−1.023	0.840
C	899	−1.561	0.917	833	−1.593	0.866
D	105	−1.724	0.834	102	−1.590	0.597
E	502	−1.944	0.546	259	−1.529	0.691
F	1848	−2.332	0.754	19	−1.348	0.802
G	41	−1.601	0.802	14	−1.220	0.715
Total	6873	−1.765	0.871	2333	−1.483	0.891

addition, there had been a marked pattern in the distribution of the root tips in the different sectors of the lysimeters, resulting from drainage of the application of the treatments, effective analysis might have been impossible. Even with the present relatively uncomplicated results, there remain some aspects of the analysis which need further examination, for example the variation in numbers of root tips with depth in the different sectors of the lysimeters.

9   Alternative designs

No analysis of the results of an experiment is complete without a review of desirable modifications to future experiments. The most obvious requirement in this particular experiment is for greater replication, if possible, of the treated and untreated lysimeters. Having only two replications of these treated and untreated lysimeters provides too few degrees of freedom for comparisons between whole lysimeters. As a result, it was necessary to use the differences between samples taken from sectors within lysimeters treated alike as estimates, possibly invalid, of the experimental errors. Ideally, any repeated experiment of this kind would require four or five lysimeters for each treatment, as well as for the control. The obvious objection to such a recommendation, however, is that it would greatly increase the amount of work, not only in the application of controlled or acidified simulated rain, but also in the counting of root tips in the sample cores from each lysimeter. Even with the present limited experiment, it was found impossible to analyse more than one core from each of the six sectors in each lysimeter.

A possible alternative, though admittedly likely to cause practical difficulties, would be to divide each lysimeter into two, by placing an impermeable barrier across its diameter. The control and acid treatment could then be applied at random to one half of each lysimeter, thus providing a simple split-plot experiment, with repeated samples within each of, say, three sectors within each half of the split lysimeter. Such an arrangement would certainly complicate the application of the treatment, as care would need to be taken to see that none of the acidified rainfall spilled into the untreated half, quite apart from the difficulty of ensuring an impermeable barrier between the soils of the two halves. Nevertheless, this split plot would probably be the most satisfactory arrangement if, as is almost certain, only a limited number of lysimeters could be provided. Arrangements would have to be made to ensure that any collected solutes from the soils were separated at the base of the lysimeter. The experimental difficulties are perhaps formidable, but not necessarily impossible.

A further important question is related to the effectiveness of dividing the lysimeters into sectors, and taking samples from each of the sectors. This restriction imposed upon the randomisation ensures a relatively even distribution throughout the lysimeters, but also complicates the subsequent analysis. There is little evidence in the present data to suggest that much advantage has been gained from the division of the lysimeters into sectors, and an equivalent number of completely random samples or cores taken from the lysimeters would probably have been as effective. Nevertheless, the control implied by the division into sectors might have been useful if there had been detectable effects due to drainage or the application of treatments because of the orientation of the lysimeters. On the whole, the practice that was adopted in the present experiment was probably desirable, and would have been even more valuable if both cores from each sector had subsequently been assessed.

Finally, the shape of the curves of the distribution of numbers of root tips with depth suggests that it would have been sufficient to record numbers of root tips in 6 cm horizons rather than in the more detailed 3 cm horizons. The reduced number of horizons would still have given six points along a curve, probably sufficient to determine an empirical relationship between the number of root tips and depth. Indeed, it would have been preferable to have had records from both cores and from each sector for the smaller number of horizons, rather than the 11 horizons for only one core from each sector.

10   Discussion

The experiment described in this chapter highlights a number of the problems which regularly beset the working statistician. Although most experimenters recognise that it is desirable to consult a statistician before the experiment is committed to a particular design, for a variety of reasons most experiments are still designed without the benefit of such advice. Either a statistician is not available at the right time, or it is felt that the advice of a statistician is likely to complicate the research to such an extent that experimenters will avoid seeking it. There is a quite strongly held view among experimenters that statisticians always ask for more replication than can be provided, and hence jeopardise the research by suggesting that it is not worth doing unless sufficient replication can be provided. There is, of course, some truth in this allegation, and, equally, some truth in the view that, unless an experiment can be done with adequate replication, and with due regard to the size of the difference which it is important to be able to detect, the research may indeed not be worth doing.

Having obtained the results of the experiment, most experimenters will then look for ways to analyse the data, and only, at this stage, recognise that some help is needed from a qualified statistician. Somewhat reluctantly, the experimenter consults the statistician, only to be told what he knew in the first place, i.e. that the difficulties of the analysis are largely due to the original design. Sometimes, however, the problems of analysis lie in an incorrect perception of the form of the appropriate analysis. In the absence of the fairly extensive statistical training which is often necessary for the analysis of even apparently simple types of data, most experimenters will hunt for an appropriate form of analysis in a textbook, or, more recently, at the nearest computing centre.

Indeed, the introduction of the electronic computer as the by now almost ubiquitous method of analysing data has, if anything, made matters worse. A vast range of analytical procedures have now been programmed, and are available, through a variety of statistical packages, to any user willing to master the somewhat daunting language in which the program description is written, including the instructions for preparing the data for input to the computer. It is now desperately easy for wholly inappropriate techniques to be used in the analysis of data, simply because the methods are available and accessible through a local computing centre. With the advent of desk-top microprocessors, complete with extensive packages of analytical techniques on disks and tapes, a high proportion of the data collected in scientific research never receives any adequate consideration or analysis. The experimenters will process their own data unless they recognise that there is a problem, or unless the results of the analysis are sufficiently bizarre for them to be alerted to the deficiencies of the analytical technique. Statisticians have not yet found an answer to this problem, but it is one which needs our urgent consideration if a great deal of valuable research is not to be wasted, not only by inadequate design, but also by inappropriate, and sometimes misleading, analysis.

Finally, it is perhaps worth stressing that, even with the use of computers, the adequate analysis of experimental data is an extremely time-consuming occupation. As a simple rule of thumb, a well-designed experiment will require approximately 10 times the amount of time for the analysis as was taken in the actual collection of the data. If the experiment was badly designed, the time for analysis may increase to 30 or even 100 times the time taken to actually collect the data. The value of good design is therefore very considerable, and emphasises the importance of early consultation with the statistician when the research is being planned.

The following abbreviated checklist asks some of the most important questions that any experimenter needs to answer. A fuller checklist is available in Jeffers (1978).

1 Have you stated clearly and explicitly the objectives of the experiment and the reasons for undertaking it?
2 Have you defined carefully the population about which you are seeking to make inferences from the results of the experiment?
3 Are the experimental treatments sufficiently well defined for the experiment to be repeated, and are they realistic?
4 Have you calculated the numbers of replications necessary to give the desired precision to the estimates or tests of significance derived from the experiment?
5 Does your choice of experimental design allow for the meaningful interpretation of the results?
6 Have the treatments and controls been allocated to the plots of the experiment by an explicit randomising procedure?
7 Are the measurements to be made in the experiment meaningful and relevant to the objectives of the experiment?
8 Are you aware of the problems that may be encountered during the analysis of the results of the experiment?

Acknowledgements

The author thanks Dr J. Dighton of the Institute of Terrestrial Ecology for permission to quote his data. This work was performed under CEGB contract RK 4166.

References

Cochran, W. G. and Cox, G. M. (1950) *Experimental designs*. New York: John Wiley & Sons.
Dighton, J., Skeffington, R. A. and Brown, K. A. (1986) Effects of simulated acid rain (pH 3) on roots and mycorrhizas of *Pinus sylvestris*. In: *Proceedings of the 1st European Mycorrhizal Congress, Dijon V*, edited by Gianazzu-Pearson.
Jacobson, J. S. (1984) Effect of acidic aerosol, fog, mist and rain on crops and trees. *Phil. Trans. R. Soc. Lond.* B, **305**, 327–38.
Jeffers, J. N. R. (1978) Design of experiments. (*Statistical checklist 1*). Grange-over-Sands: Institute of Terrestrial Ecology.
Jeffers, J. N. R. (1979) Sampling. (*Statistical checklist 2*). Grange-over-Sands: Institute of Terrestrial Ecology.
Jeffers, J. N. R. (1980) Modelling. (*Statistical checklist 3*). Grange-over-Sands: Institute of Terrestrial Ecology.
McNeil, D. R. (1977) *Interactive data analysis*. New York: John Wiley & Sons.
Snedecor, G. W. (1946) *Statistical methods*. Iowa: Iowa State College Press.
Tukey, J. W. (1977) *Exploratory data analysis*. London: Addison-Wesley.

8

On identifying yeasts and related problems

J. C. GOWER AND R. W. PAYNE

1   Introduction

It was in 1960 that J.C.G.† initially became interested in problems
of classification in the sense of *constructing* classes of individuals. This
type of problem contrasts with the use of the term 'classification' in
classical statistics which is concerned with discrimination, that is the
problem of *assigning* an individual to one of several *a priori* classes.
Sneath (1957) had written a paper in the *Journal of General Microbiology*
about a computer program that he had developed for generating a hierarchical classification of bacterial strains. His approach was to compute
coefficients (the simple matching and related coefficients) that gave a
measure of the similarity between each pair of strains as judged by their
abilities, either 'present' or 'absent', to thrive on a selection of nutritive
bases. Sneath used a simple hierarchical classification algorithm, now
termed the single-linkage algorithm, in which strains are successively
grouped, the pair of groups fusing at each stage which have the biggest
similarity between any pair of strains, one from each group. Similar work
was reported from the USA by Sokal and Michener (1958) (classifying
bees) and Rogers and Tanimoto (1960) (botanical classification) while
Williams and Lambert (1959), in Southampton, were concerned with the
ecological classification of quadrats based on the presence/absence of
plant species within quadrats sampled from a region. It was the newly
available computers that stimulated these and other scientists to investigate
problems that had not previously been viewed as amenable to a numerical
approach. It is notable that none of this early work was done by
professional statisticians or computer scientists. Indeed statisticians were

† To avoid circumlocutions, when it is necessary to distinguish between the
  authors, initials will be used. R.W.P. became involved in 1972, on joining
  Rothamsted.

slow to take an interest in this area, and in some cases reacted as if the fact that the word 'classification' was not being used in its statistical sense of discrimination implied inadequacy in the work (see, for example, the discussion following Cormack (1971)). Of course the subject has grown enormously since 1971 but much still remains to be done to attain a proper theoretical treatment of the many problems involved: see Jardine and Sibson (1971) for a beginning. This development is an excellent example of how theoretical statistics often develops from an applied stimulus rather than vice versa but it is not the main theme of the present paper.

The problem that initiated J.C.G.'s interest in classification came from Dr Margaret Pleasance, a scientist at the Agricultural Research Council's Low Temperature Research Station (LTRS) in Cambridge. The LTRS later split into two institutes, the Food Research Institute (FRI) in Norwich and the Meat Research Institute (MRI) located in Langford near Bristol. In 1985 these two institutes, together with part of the National Institute for Research in Dairying, near Reading, merged to form a single multi-site Institute of Food Research. Thus the 'low temperatures' of interest to the LTRS were those associated with food storage, and bacterial activity was a major concern. Dr Pleasance had read the paper by Sneath and wondered whether similar methods could help her in the classification of a major bacterial group. Accordingly J.C.G. set to work to write a suitable computer program (in machine code) and this led to an interest in coefficients of similarity, algorithms for hierarchical classification and in the scientific underpinning of classification. Many applications followed in collaboration with many scientists, often taxonomists. The starred items in the list of references give an indication of the diversity of the organisms studied. The original programs have been steadily developed by G. J. S. Ross and now form a package named Clasp (Classification Program) whose main features have also been incorporated into Genstat.

Classification is only one of three major concerns of taxonomists, the others being (i) evolution and (ii) how to identify species. There is a belief that if the course of evolution were accurately known then this would supply a tree giving the best (usually referred to as the most natural) hierarchical classification which would necessarily offer the best basis for identification. Whether there is much truth in this belief is unsubstantiated but it cannot be denied that the course of evolution is not known accurately for any group of animals or plants and that only the sketchiest information is usually available. Indeed because of this paucity of information the argument is commonly reversed and what is deemed to be the *best* classification strongly influences notions of evolution. Perhaps part of the difficulty is the predilection, very strongly influenced by Linnaeus, that taxonomists have for hierarchical classifications. Because evolution too is usually thought of, at least in broad terms, as a hierarchical process representable by a tree structure, there is a tendency to confuse the two different problems.

What is meant by *best* in the above discussion is controversial. There are two major divisions among taxonomists: the Cladists, who try to build up evolutionary trees directly by minimising some measure of the number of character changes as one moves between adjacent nodes of a putative evolutionary tree; and the Pheneticists, who base hierarchical classifications on measures of overall similarity (i.e. similarity coefficients). This division is reinforced by considerations of the different types of characteristics admitted; whether they have basic genetic content such as DNA or amino acid sequences that might be thought of as specially likely to contain evolutionary information, or whether phenetic characters, traditionally used by taxonomists, that are mainly concerned with (visual) resemblance are to be used. For a recent discussion of the differing views see the various papers in Felsenstein (1983). Thus *best* may be defined in terms either of minimising measures of evolutionary change or of maximising measures of resemblance as approximated by tree structures. A third approach is to separate taxonomy from evolutionary speculation by regarding it as an information system in which information on species can be retrieved efficiently. This approach is favoured by Gilmour (1937) who sums it up in the statement *that a system of classification is the more natural the more propositions there are that can be made regarding its constituent classes.* Gower (1974) gave a mathematical formalisation, termed *maximal predictive classification*, in which each non-hierarchic class is characterised by a *class predictor* which indicates the state of each character displayed by the majority of the species in the class. It is perhaps no surprise that it is simple to assign to maximal predictive classes; one merely needs to know which class predictor corresponds most closely to the specimen being examined. This association with the identification of species is an important general aspect of the informational approach, which taken to its conclusion gives yet another definition of the *best* classification as the one that makes identification most efficient. Maximal predictive classification has been implemented in Genstat and used with many sets of organisms; in particular, Barnett, Bascombe and Gower (1975) discuss maximal predictive aspects of the classification of yeast species. The relationship between classification and identification is discussed further by Gower (1973, 1975) and the development of methods for identification is discussed in the following.

2   Identification keys

Whereas statistical discriminant analysis is concerned with assigning a sample to one of a few, often two, populations in the presence of uncertainty, taxonomists are more concerned with assigning to one of several hundred populations when uncertainty often plays a minor role, because qualititative characteristics can be found that vary little, or not at all, within all or some populations. Thus a methodology has to be worked out for general identification problems of this latter kind.

Taxonomists have been concerned with this problem for several hundred years and nearly every botanical handbook contains a key to help identify plants. An identification (or diagnostic) key has a tree-structure where each node corresponds to a diagnostic question 'Which one of a named set of attributes does the specimen to be identified possess?' The outcome determines which branch of the tree to follow and hence the next diagnostic question and ultimately the correct identification. Often there are only two attributes, concerning the presence and absence of a particular character or the reponse to a binary test, but multi-state characters or tests are permissible.

Traditionally keys have, perforce, been constructed by hand and often embody the lifetime's experience of their designers. Very often such keys are influenced by evolutionary notions and therefore may not be the most efficient; that is a shorter or easier-to-use key may exist. There may be a need for several keys for the same group of organisms; for example a botanical key using floristic characters is useless when the specimen to be identified is not in flower. The mechanisation of the two very different processes of (i) constructing keys and (ii) the actual process of identification, made much slower progress than had that of forming classifications, perhaps because of the scepticism of traditional taxonomists. Apart from our own work, other pioneers were Hall (1970), Pankhurst (1970), Morse (1971) and Dallwitz (1974). Sneath and Sokal (1973), pp. 388–400, give an account of the situation at that time.

J.C.G. was aware of the problem early on but first considered it in detail in 1964 when Dr T. Webster of the National Institute of Agricultural Botany (NIAB) consulted him on the possibility of distinguishing between some 40 varieties of field bean. The diagnostic key approach turned out to be inappropriate because the small differences between varieties were more consistent with a set of 40 overlapping populations separable by a discriminant analysis.

For many years, J.C.G. had been consulted by Dr J. A. Barnett of FRI (and later working at the University of East Anglia) on problems concerned with yeasts, some of which simply involved standard statistical techniques while others, like the establishment of metabolic pathways based on the capacity of various species to grow on different nutrients, were more complicated. In 1968, Dr Barnett had to identify thousands of strains of yeast isolated from fruit. At that time, it was appropriate to base identification on the responses of each strain to 40 physiological tests and, because many of the tests took two weeks to complete, for any one yeast it was imperative to do all the tests simultaneously. Hence, as few tests as possible had to be used. Accordingly, Dr Barnett made a list of 68 species which, from publications, seemed to represent most kinds of yeast associated with fruits. Assuming that about 90% of his isolates would be identifiable as belonging to some of these species he asked the question 'What is the minimum number of tests needed for distinguishing between these species?' At that time, there was no simple answer to his question because there were three possible responses to each test, namely: positive, negative or query. A partial solution to his problem, that was applied to the fruit yeasts (Barnett, 1971; Buhagiar and Barnett, 1971), involved making a form of diagnostic key. This was the first step towards generating identification keys for all yeast species. By 1970 about 360 species of yeast were recognised.

Yeasts are, of course, a commercially important group of organisms in baking, brewing, wine making and in medicine, and correct identification of strains can be crucial. The initial Rothamsted programs for constructing keys, written by Bridget Lowe and Bruce Lauckner, were of limited use because output was in a coded form that was not readily interpretable. Pankhurst's program gave good readable output and was used as the basis for the first published computer generated key to the yeasts (Barnett and Pankhurst, 1974). Subsequent work involved R.W.P. in collaboration with Dr Barnett and also Dr D. Yarrow, a yeast taxonomist from the Centraal bureau voor Schimmelcultures at Delft, Netherlands.

Starting in 1972, the earlier programs were replaced by the program Genkey (Payne, 1975, 1978, 1985). This can construct probabilistic and non-probabilistic keys as well as other identification aids such as diagnostic tables and polyclaves. Probabilistic methods become necessary when a stage is reached on the tree where no diagnostic questions exist that separate species with certainty. A diagnostic table is a species × character table arranged in lexicographic order and is used to identify a specimen as one would use a dictionary. A polyclave is a set of punched cards, one for each character-state, and with a position on the cards allocated to each species. To use a polyclave see, for example, Payne, Lamacraft and White (1981). Genkey was used to construct the keys and tables in the book *A Guide to Identifying and Classifying Yeasts* (Barnett, Payne and Yarrow,

1979) and to construct and typeset those in *Yeasts: Characteristics and Identification* (Barnett, Payne and Yarrow, 1983). Genkey has been written, as far as possible, in Standard Fortran, allowing versions to be produced for seven computer ranges and it is in use in eight countries.

3 Methodology

Quite apart from their economic importance the yeasts have provided a very challenging set of data that has stimulated the development of both methodology and computer programs (Payne, 1983). The large number of species that have been described (497 at 1 November 1984) provides clear motivation for the use of computers for key construction; the size of the data set allows study by simulation, using randomly selected subsets of the species, to investigate the behaviour of various methods suggested for constructing keys (see Payne and Dixon, 1984). Most importantly, the construction of the keys and tables for identifying yeasts has not only exploited most of the known techniques but also has required new methods to be developed that are briefly described in the remainder of this section.

3.1 *Key construction*

The most efficient key is usually defined as that with the smallest expected number of tests (or characters) per identification or, where the tests have different costs, that with minimum expected cost of identification. These expectations may take count of the *a priori* probabilities of the different species or all species may be assumed equiprobable. To find such a key for any particular set of species is an N-P complete problem and so is not feasible for large sets of data, like the yeasts. Thus heuristic methods are used which construct the key sequentially, selecting first the test that 'best' divides the species into sets (set k for test i containing the species that can give result k to that test) then selecting the best test to use within each set, and so on until the sets each contain only one species and are thus at the endpoints of the branches.

The 'best' test is usually determined by some *selection criterion function*. The earliest functions considered only binary tests and allowed for neither differing costs nor variable responses (i.e. where members of the same species give different results to a test). The 'best' test was then taken to be the one that divided the species into sets of most nearly equal size. With the yeasts many responses are variable. Others are unknown – some tests may never have been recorded for some of the rarer species – and must also be treated as variable. Gower and Barnett (1971) developed a function that allowed for unknown responses to binary tests but this does not readily extend to tests with different costs nor to tests with more than two responses, as was noted by Gower and Payne (1975), who compared existing criteria and derived new ones with a better mathematical justification. Further functions with a mathematical derivation were produced by Brown (1977) but the necessary principle for incorporating differing costs was due to Dallwitz (1974) whose function, for test i, represents the expected cost of completing the identification of a specimen starting at the current point of key, assuming that test i is used next. Dallwitz assumed that the key would be completed by branches with equal lengths. Payne (1981) derived similar criteria, (i) assuming that the key is completed optimally (which, for species with unequal prior probabilities, does not necessarily generate branches with equal lengths) and (ii) for more pessimistic (or realistic) assumptions; it was also shown that different functions were appropriate for variable and for unknown responses. Payne and Dixon (1984) performed simulations to study the behaviour of some of these criteria, using both the yeasts and artificial data, and found that no criterion was uniformly better than the others. This result supports the strategy adopted in Genkey, which offers several criteria among which users may choose.

3.2   *Irredundant test sets*

Many of the tests involved in the identification of yeasts take up to 14 days to complete. Hence, it is usual to do all the tests that occur in the key, simultaneously, before the key is used (instead of doing only those tests required sequentially on the particular branches taken through the key). Thus the total number, or total cost, of all the different tests in the key becomes the main measure of its efficiency. A set of tests that contains no redundant tests, i.e. tests that can be omitted without causing any pair of species to become indistinguishable, is termed *irredundant*. Clearly both the set of tests of minimum size and the set with minimum (total) cost are irredundant, since otherwise a better set could be obtained by deleting one (or more) redundant tests.

An algorithm had been devised, in several different contexts, for constructing all irredundant sets. Full details and the original references are given in the review paper by Payne and Preece (1980). The implementation in Genkey contains various short cuts, devised by Willcox and Lepage (1972) and Payne and Preece (1980), to improve efficiency.

Irredundant test sets can also be devised to allow a particular species to be distinguished from all the other species and all such sets for a particular species can be constructed by an adaptation of the same algorithm. However, for some of the species in Barnett, Payne and Yarrow

(1983), there were so many sets that it was not feasible to consider them all. When this occurs, sequential algorithms are generally used, in which tests are selected one at a time until all the required pairs of species can be distinguished. Improved criteria for selecting these tests were devised and modifications were made to allow users a choice of several different sets, to cater for different preferences for tests, as had also been done by Barnett, Payne and Yarrow (1979), using the original (non-sequential) algorithm.

Fig. 1. One of the smaller keys from Barnett, Payne and Yarrow (1983)

Key No. 12: Yeasts that utilize methanol (test no. 52 positive)

Key involving physiological tests only

Yeasts in Key No. 2
37 *Candida boidinii*
46 *Candida cariosilignicola*
59 *Candida entomophila*
102 *Candida maris*
108 *Candida methanosorbosa*
109 *Candida methylica*
116 *Candida nemodendra*
117 *Candida nitratophila*
127 *Candida pignaliae*
128 *Candida pinus*
149 *Candida sonorensis*
155 *Candida succiphila*
246 *Hansenula capsulata*
250 *Hansenula glucozyma*

251 *Hansenula henricii*
255 *Hansenula minuta*
258 *Hansenula nonfermentans*
259 *Hansenula ofunaensis*
261 *Hansenula philodendra*
262 *Hansenula polymorpha*
267 *Hansenula wickerhamii*
337 *Pichia kodamae*
338 *Pichia lindneri*
341 *Pichia methanolica*
346 *Pichia naganishii*
353 *Pichia pastoris*
356 *Pichia pini*
371 *Pichia trehalophila*

Tests in Key No. 12
1 D-Glucose fermentation
16 D-Glucosamine growth
21 L-Rhamnose growth
22 Sucrose growth
27 Salicin growth
36 Erythritol growth
39 L-Arabinitol growth
42 Galactitol growth
54 Nitrate growth
77 0.1% Cycloheximide growth

number of different tests 10

Key No. 12

	Negative	Positive
1 Nitrate growth	2	17
2 Erythritol growth	3	8
3 Salicin growth	4	6
4 Galactitol growth	5	*Hansenula ofunaensis*
5 D-Glucose fermentation	*Candida maris*	*Pichia pastoris*
6 L-Arabinitol growth	*Hansenula nonfermentans* *Pichia lindneri*	7
7 L-Rhamnose growth	*Candida sonorensis*	*Pichia lindneri*
8 Galactitol growth	9	14
9 Sucrose growth	10	13
10 Salicin growth	11	12
11 0.1% Cycloheximide growth	*Candida pinus*	*Pichia trehalophila*
12 0.1% Cycloheximide growth	*Pichia kodamae* *Pichia pini*	*Pichia methanolica* *Pichia pini*
13 L-Rhamnose growth	*Candida entomophila*	*Pichia naganishii*
14 D-Glucose fermentation	*Candida nemodendra*	15
15 D-Glucosamine growth	16	*Candida succiphila*
16 0.1% Cycloheximide growth	*Pichia kodamae*	*Pichia methanolica*
17 Erythritol growth	18	26
18 D-Glucosamine growth	19	24
19 L-Arabinitol growth	20	21
20 0.1% Cycloheximide growth	*Hansenula henricii*	*Hansenula nonfermentans*
21 L-Rhamnose growth	22	23
22 Salicin growth	*Candida pignaliae*	*Hansenula minuta*
23 0.1% Cycloheximide growth	*Hansenula henricii*	*Hansenula glucozyma*
24 Galactitol growth	25	*Hansenula ofunaensis*
25 Salicin growth	*Candida nitratophila*	*Candida methanosorbosa*
26 Sucrose growth	27	33
27 L-Rhamnose growth	28	30
28 D-Glucose fermentation	*Hansenula philodendra*	29
29 Salicin growth	*Candida boidinii*	*Hansenula capsulata*
30 D-Glucosamine growth	31	*Hansenula capsulata*
31 Salicin growth	32	23
32 D-Glucose fermentation	*Hansenula wickerhamii*	*Candida methylica*
33 D-Glucosamine growth	*Hansenula polymorpha*	*Candida cariosilignicola*

3.3   *Printing the keys*

For publication it is important to make the printed keys as short as possible. Subject to constraining the tests to be used to the chosen irredundant set as explained in Section 3.2, it is still useful to construct an efficient key, as described in Section 3.1, as this tends to decrease the length of the printed key. Further savings can be made by using one of the compact representations of Payne, Walton and Barnett (1974), as shown in Figure 1, also by the use of *reticulation* (Payne, 1977), which allows duplicate sections, that can arise in a printed key when responses are variable, to be printed only once. Even so, the 18 keys presented by Barnett, Payne and Yarrow (1983) occupy 90 pages.

3.4   *Keys to the genera*

A major characteristic used for distinguishing between yeast genera is the mode of sexual reproduction rather than the nutritional characteristics that are most convenient for identification. Consequently many of these nutritional characteristics vary over the species within the genera and there are many pairs of genera for which there is no one test that can distinguish the species of the first genus from those of the second. Thus keys to identify genera cannot be constructed in the same way as those for species. Similar problems can also occur with groupings defined by taking a particular level of a hierarchical classification. Payne, Yarrow and Barnett (1982) showed how to adapt the methods of Section 3.2 to such situations.

3.5   *Computer-based identification programs*

Identification keys have the disadvantage that the user is generally given no choice as to which tests to use (see, for example, Payne and Preece, 1980 and Payne, 1980). Also, with most keys, an error in applying or observing any of the tests will lead to an incorrect identification (although Payne and Preece (1977) describe how extra tests can be incorporated to protect against this or how to construct subsidiary keys to check the identification obtained from the main key). These disadvantages can be overcome by computer-based identification systems that incorporate test-selection algorithms similar to those in key construction programs.

The *Yeast Identification Program* of Barnett, Payne and Yarrow (1985) lists the yeast species consistent with any observed set of test results; if there is more than one such species, a set of tests can be selected to complete the identification. Errors in results already observed can be allowed for and extra tests can be selected to guard against later errors. The program also enables users to list yeasts that have a specified set of characteristics as may be desired, for example, in industrial applications.

4   Conclusion

Perhaps the main interest in this report is that it records collaboration between scientists and statisticians that has covered a period of well over 20 years. This is by no means atypical, as effective consulting calls for a statistician who has a good understanding of the field of application and this can only be built up over time. Not only does background knowledge aid initial communication between statistician and client but it also reduces misunderstandings and helps the statistician to recognise (a) areas where he might unsuspectedly be able to make a contribution and (b) difficulties that might not surface otherwise. The impression given by some textbooks in applied statistics, that a statistician armed solely with a battery of techniques can be an effective consultant in almost any field of application, is nonsense. The interplay between statistical ideas and those provided by clients, in our case taxonomists, is a stimulus that leads to fruitful and useful work for both parties. Unusually for a statistical problem, probability has played little part in the work discussed. Indeed some would say that therefore the work is not statistical. We would not agree with this view, holding that the essential feature of a statistical problem is that it be concerned with the analysis of data. In our opinion some statisticians are too rigid in what they accept as a valid statistical problem. It is our clients who have the problems and it is our job as statisticians to do what we can to help without worrying too much about demarcation disputes. In taxonomy the line between probabilistic and non-probabilistic models is a fine one. Taxonomists concerned with identification will, very sensibly, proceed as far as they can by using characteristics that are constant within species. As a last resort, to separate out overlapping species, probabilistic methods must be used and this is the strategy that we have followed.

Thus starting with some experience of taxonomic problems, an enquiry about constructing taxonomic keys has led not only to providing software to cover the initial problem but also to the study of several methodological problems whose solutions, as evidenced by the references, have made the process of identification more efficient, more reliable and more flexible. The bulk and detailed information of identification keys has naturally led to a study of the technological problems of computer-aided book printing. With the information already in computer-readable form, it is now possible to typeset keys directly (Payne, 1984) and even to generate and automatically typeset descriptions of the species (Barnett, Payne and Yarrow, 1983). This avoids the possibility of transcription errors, which in a key would be disastrous, and also very much reduces production costs.

The development of the project has involved much reading and we have thus become aware of related work in coding theory, computer science, pattern recognition, the theory of questionnaires, psychology, search theory, etc.; see the review of Payne and Preece (1980). We note the close links with medical diagnosis, especially the approach nowadays associated with work on expert systems. We hope that our work may be of use to those working in these other areas, to other statisticians who are asked about similar problems and to all those concerned with identification.

References

Barnett, J. A. (1971) Selection of tests for identifying yeasts. *Nature Biol., Lond.*, **232**, 221–2.

Barnett, J. A., Bascombe, S. and Gower, J. C. (1975) A maximal predictive classification of Klebsielleae and of the yeasts. *Journal of General Microbiology*, **86**, 93–102.

Barnett, J. A. and Pankhurst, R. J. (1974) *A New Key to the Yeasts: A Key for Identifying Yeasts based on Physiological Tests Only*, Amsterdam: North-Holland, 273pp.

Barnett, J. A., Payne, R. W. and Yarrow, D. (1979) *A Guide to Identifying and Classifying Yeasts*, Cambridge University Press, 315pp.

Barnett, J. A., Payne, R. W. and Yarrow, D. (1983) *Yeasts: Characteristics and Identification*, Cambridge University Press, 811pp.

Barnett, J. A., Payne, R. W. and Yarrow, D. (1985) *Yeast Identification Program*, Cambridge University Press, 40pp.

Brown, P. J. (1977) Functions for selecting tests in diagnostic key construction. *Biometrikia*, **64**, 589–96.

Buhagiar, R. W. M. and Barnett, J. A. (1971) The yeasts of strawberries. *J.appl.Bact.*, **34**, 727–39.

Cormack, R. M. (1971) A review of classification. *J.R.Statist.Soc.* A, **134**, 321–67.

Dallwitz, M. J. (1974) A flexible computer program for generating identification keys. *Syst.Zool.*, **23**, 50–7.

*Eddy, B. P. and Carpenter, K. P. (1964) Further studies on Aeromonas. II. Taxonomy of Aeromonas and C27 strains. *J.Appl.Bact.*, **27**, 96–109.

Felsenstein, J. (1983) Statistical inference of phylogenies, *J.R.Statist.Soc.* A, **146** (3), 246–72.

Gilmour, J. S. L. (1937) A taxonomic problem. *Nature*, **134**, 1040–2.

Gower, J. C. (1973) Classification problems. *Bull.Int.Inst.Statist.* **45**, 471–7.

Gower, J. C. (1974) Maximal predictive classification. *Biometrics*, **30**, 643–54.

Gower, J. C. (1975) Relating classification to identification. In: *Biological Identification with Computers* (ed. R. J. Pankhurst), London: Academic Press, pp. 251–63.

Gower, J. C. and Barnett, J. A. (1971) Selecting tests in diagnostic keys with unknown responses. *Nature*, **232**, 491–3.

Gower, J. C. and Payne, R. W. (1975) A comparison of different criteria for selecting binary tests in diagnostic keys. *Biometrika*, **62**, 665–75.

Hall, A. V. (1970) A computer-based system for forming identification keys. *Taxon*, **19**, 12–18.

Jardine, N. and Sibson, R. (1971) *Mathematical Taxonomy*, New York: Wiley, 286pp.

Morse, L. E. (1971) Specimen identification and key construction with
    time-sharing computers. *Taxon*, **20**, 269–82.
Pankhurst, R. J. (1970) A computer program for generating diagnostic keys.
    Computer J., **13**, 145–51.
Payne, R. W. (1975) Genkey: a program for constructing diagnostic keys. In:
    Biological Identification with Computers (ed. R. J. Pankhurst), London:
    Academic Press, pp. 65–72.
Payne, R. W. (1977) Reticulation and other methods of reducing the size of
    printed diagnostic keys. *Journal of General Microbiology*, **98**, 595–7.
Payne, R. W. (1978) Genkey – a program for constructing and printing
    identification keys and diagnostic tables, Harpenden: Rothamsted
    Experimental Station.
Payne, R. W. (1980) Alternative characters in identification keys. *Classification
    Society Bulletin*, **4**, 16–21.
Payne, R. W. (1981) Selection criteria for the construction of efficient
    diagnostic keys. *Journal of Statistical Planning and Inference*, **5**, 27–36.
Payne, R. W. (1983) Construction and automatic typesetting of keys to the
    yeasts. In: *Groupe de Recherche, Claude Francois Picard, Structures de
    l'Information, publication No. 34: Journées Tourangelles Utilisation de
    l'Information, des Questionnaires et des Ensembles Flous dans les Problems
    Décisionnels*, pp. 1–113. Paris: Groupe de Recherche 22, Université Pierre et
    Marie Curie.
Payne, R. W. (1984) Computer construction and typesetting of identification
    keys. *New Phytologist*, **96**, 631–4.
Payne, R. W. (1985) Genkey: a general program for constructing aids to
    identification. *Informatique et Sciences Humaines*, **64**, 45–62.
Payne, R. W. and Dixon, T. J. (1984) A study of selection criteria for
    constructing identification keys, *COMPSTAT 1984: Proceedings in
    Computational Statistics* (eds. T. Havranek, Z. Sidak and M. Novak),
    Vienna: Physica-Verlag, pp. 148–53.
Payne, R. W., Lamacraft, R. R. and White, R. P. (1981) The dual polyclave:
    an aid to more efficient identification, *New Phytologist*, **87**, 121–6.
Payne, R. W. and Preece, D. A. (1977) Incorporating checks against observer
    error into identification keys, *New Phytologist*, **79**, 201–7.
Payne, R. W. and Preece, D. A. (1980) Identification keys in diagnostic tables:
    A review. *J.R.Statist.Soc.* A, **143**, 253–92.
Payne, R. W., Walton, E. and Barnett, J. A. (1974) A new way of representing
    diagnostic keys. *Journal of General Microbiology*, **83**, 413–14.
Payne, R. W., Yarrow, D. and Barnett, J. A. (1982) The construction by
    computer of a diagnostic key to the genera of yeasts and other such groups
    of taxa. *Journal of General Microbiology*, **128**, 1265–77.
*Rayner, J. H. (1965) Multivariate analysis of montmorillonite. *Clay Minerals*,
    6, 59–70.
*Rayner, J. H. (1966) Classification of soils by numerical methods. *J.Soil Sci.*,
    17, 79–92.
Rogers, D. J. and Tanimoto, T. T. (1960) A computer program for classifying
    plants. *Science*, **132**, 1115–18.
*Sheals, J. G. (1964) The application of computer techniques to Acarine
    taxonomy: a preliminary examination with species of the
    Hypoaspis-Androlaelaps complex (Acarina) *Proc.Linn.Soc.Lond.*, **176**, 11–21.
*Smith, I. W. (1963) The classification of 'bacterium salmonicida'. *Journal of
    General Microbiology*, **53**, 263–74.

Sneath, P. H. A. (1957) The application of computers to taxonomy. *Journal of General Microbiology*, **17**, 201–26.

Sneath, P. H. A. and Sokal, R. R. (1973) *Numerical Taxonomy: The Principles and Practice of Numerical Classification*, San Francisco: Freeman, 573pp.

Sokal, R. R. and Michener, C. D. (1958) A statistical method for evaluating systematic relationships. *Univ.Kansas Sci.Bull.*, **38**, 1409–38.

Willcox, W. R. and Lapage, S. P. (1972) Automatic construction of diagnostic tables. *Computer J.*, **15**, 263–7.

Williams, W. T. and Lambert, J. M. (1959) Multivariate methods in plant ecology. I. Association-analysis in plant communities. *J.Ecol.*, **47**, 83–101.

9

Uneven sex ratios in the light-brown apple moth: a problem in outlier allocation

TOBY LEWIS

One of the phenomena investigated by P. W. Geier and co-workers at the CSIRO Division of Entomology, Canberra, in a series of studies (Geier and Springett (1976), Geier and Briese (1977), Geier and Oswald (1977)) of the light-brown apple moth, *Epiphyas postvittana* (Walker), was the tendency for males to be consistently less abundant than females in samples drawn from Australian field populations. The existence of a heritable condition ('Q-condition') in 'carrier' females of the light-brown apple moth, which caused them to produce predominantly female progenies, was established by Geier and Briese (1977). A further extensive investigation of the Q-condition was carried out by Geier and Briese in 1978. In this connection they consulted me (at the time a visiting member of the CSIRO Division of Mathematics and Statistics at Canberra) with regard to the modelling of some aspects of their data, and I give an account here of the statistical work (published as part of Geier, Briese and Lewis (1978)) which I carried out in support of their research. Apart from its statistical content, it perhaps has some interest as a cautionary tale about the pitfalls which a statistician may encounter in offering too dogmatic an interpretation of a researcher's results.

1 The problem

The problem was outlined as follows in a letter to me from Dr D. T. Briese, quoted here with his permission:

> ...As you can see in Fig. 1a, the distribution of the percentage males per sample in the laboratory stock appears simple enough. I have assumed it would be binomial with a [*p*-value] of 0.511 males. However, we recently reared progeny from individual females collected in the field over a wide part of South-eastern Australia and Tasmania. These contain some females which produce either no males or a very few males in their progeny.

(Parthenogenesis does not occur in this species, and mating is obligatory for reproduction.) The...distribution of sex ratio in these field samples is shown in Figs 1b and 1c.... Considered overall the distribution appears to be a combination of two

Fig. 1. Sex ratios in LBAM (light-brown apple moth) as percentage of males per sample

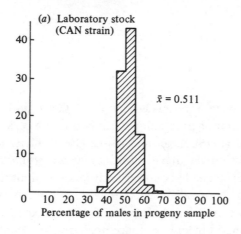

(a) Laboratory stock (CAN strain)

$\bar{x} = 0.511$

Percentage of males in progeny sample

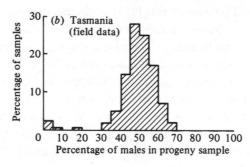

(b) Tasmania (field data)

Percentage of samples

Percentage of males in progeny sample

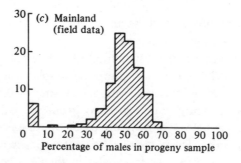

(c) Mainland (field data)

Percentage of males in progeny sample

binomial ones. If individual locations are considered, ...the basic pattern of the distribution† appears to be the same, although the relative frequencies of the peaks vary.

My main concern is whether there is a frequency distribution that can describe the overall field data, and whether one can consider there to be a range of this type of distribution which would fit the sex ratio data of individual localities....

A secondary problem arises with the actual frequency distribution of the percentage of 'Q' type females (the...females [producing few or no males]) in individual samples†...Is there a type of frequency distribution which could describe this?...

In statistical terms we can say that, in the population studied, field-collected adult females of the light-brown apple moth produced progenies whose sizes (n) were mostly in the range 30–70. If adult female i ($i = 1, 2, ...$) produced progeny of size n_i consisting of m_i males and $f_i = n_i - m_i$ females, the proportion of males m_i/n_i is the (observed) *sex ratio*. We can denote it by $\hat{\theta}_i$ since it estimates a *sex ratio parameter* θ_i associated with parent female i. The great majority of the progenies in the study ('*normal-type*' *progenies*) contained comparable numbers of male and female offspring, implying θ-values at or near $\frac{1}{2}$. By contrast, about 5% of the progenies ('*Q-type*' *progenies*) had few or no males, and implied very small θ-values. Essentially what the entomologists wanted was a satisfactory probability model for these data.

2   The data

The full set of data consisted of the numbers of males and females in progenies of 439 field-collected female moths from 68 different localities in Australia. By way of illustration, the data for four of these 68 localities, Murray Bridge, Tailem Bend, Muswellbrook and Sale, are set out in Table 1, comprising 60 of the 439 progenies. These had been chosen by Dr Briese, with pooling of the data from the neighbouring localities Murray Bridge and Tailem Bend, as '...two of the samples containing a relatively high proportion of "Q" type females, plus one (Sale) which is apparently made up only of normal females'.

By inspection, 413 of the 439 progenies had been allocated initially to the normal-type group, as having sex ratio values m_j/n_j consistent with a binomial $B(n_j, \theta)$ model for m_j with $\theta \approx \frac{1}{2}$; for example, in Table 1,

---

† Dr Briese gave histograms of the Murray Bridge/Tailem Bend and Muswellbrook data listed below in Table 1, and of the proportions Q_i/T_i of Q-type females from the data listed in Table 4. These histograms are not reproduced here.

progenies $p = 3$ to $p = 7$ of the Murray Bridge set were so allocated, as also progenies $p = 7$ to $p = 12$ of the Tailem Bend set, progenies $p = 7$ to $p = 22$ of the Muswellbrook set, and all the Sale progenies. This left 26 progenies which were allocated initially to the Q-type group. The frequency distribution of the 413 sex ratios m_j/n_j ($j = 1, \ldots, 413$) is given in grouped form in Table 2. (In the analysis, the ungrouped exact values were used.) Note that a binomial distribution could not be expected to model these data exactly, since the sex ratios are based on a range of different progeny sizes n_j; see distribution of progeny sizes on page 125.

Table 1. *The data for four of the 68 localities*

Progeny no.	Locality											
	Murray Bridge, South Australia (locality no. 15) $T_{15} = 7$			Tailem Bend, South Australia (locality no. 16) $T_{16} = 12$			Muswellbrook, New South Wales (locality no. 38) $T_{38} = 22$			Sale, Victoria (locality no. 27) $T_{27} = 19$		
	m_i	f_i	m_i/n_i	m_i	f_i	m_i/n_i	m_i	f_i	m_i/n_i	m_i	f_i	m_i/n_i
$p = 1$	0	21	0	0	60	0	0	61	0	21	36	0.368
2	0	20	0	0	51	0	0	48	0	6	9	0.400
3	26	32	0.448	0	14	0	0	42	0	26	35	0.426
4	29	35	0.453	1	54	0.018	0	35	0	22	27	0.449
5	25	30	0.455	2	53	0.036	1	42	0.023	26	31	0.456
6	18	18	0.500	7	47	0.130	1	29	0.033	24	26	0.480
7	34	20	0.630	3	9	0.250	11	18	0.379	28	28	0.500
8				14	26	0.350	22	31	0.415	33	33	0.500
9				16	23	0.410	27	34	0.443	24	24	0.500
10				24	24	0.500	29	33	0.468	31	30	0.508
11				39	31	0.557	24	27	0.471	29	27	0.518
12				40	22	0.645	29	33	0.468	34	29	0.540
13							25	28	0.472	35	28	0.556
14							26	23	0.531	26	20	0.565
15							38	33	0.535	31	22	0.585
16							14	12	0.538	26	17	0.605
17							23	19	0.548	30	19	0.612
18							31	25	0.554	13	8	0.619
19							20	14	0.588	36	22	0.621
20							6	4	0.600			
21							34	22	0.607			
22							12	7	0.632			

T_l number of field-collected adult female moths from locality l
n_i size of progeny i; m_i, f_i numbers of male and female offspring in this progeny ($m_i + f_i = n_i$); m_i/n_i = sex ratio
——— suggested boundary between Q-type and normal-type progenies (entomologists' initial classification)

Range of values of n_j	Frequency
6–10	6
11–20	18
21–30	21
31–40	51
41–50	53
51–60	160
61–70	98
71–77	6
	413

Table 2. *Frequency distribution of 413 sex ratios initially classified as normal*

Value of $\theta_j = m_j/n_j$ (to nearest 0.01)	Frequency	Value of $\theta_j = m_j/n_j$ (to nearest 0.01)	Frequency
0.17	1 ($m = 1, n = 6$)		
0.25	1 ($m = 3, n = 12$)		
0.31	1	0.51	11
0.32	0	0.52	28
0.33	3	0.53	25
0.34	2	0.54	24
0.35	4	0.55	24
0.36	3	0.56	13
0.37	3	0.57	15
0.38	6	0.58	12
0.39	4	0.59	7
0.40	3	0.60	9
0.41	10	0.61	14
0.42	14	0.62	11
0.43	10	0.63	4
0.44	12	0.64	3
0.45	21	0.65	2
0.46	14	0.66	0
0.47	24	0.67	3
0.48	23	0.68	2
0.49	15	0.69	2
0.50	29		
		0.75	1 ($m = 9, n = 12$)
			413

Details of the remaining 26 sex ratios, which because of their low values were allocated provisionally to the Q-type group, are listed in Table 3. Finally, Table 4 shows the distribution over the 68 localities l ($l = 1, ..., 68$) of the total number T_l of field-collected adult female moths, the number Q_l of these which were initially classified as Q-type, and the number N_l which were initially classified as normal-type ($Q_l + N_l = T_l$).

3   Statistical analysis

Taking the numbers of males m_i in the 439 progenies studied to be independent observations from binomial distributions $B(n_i, \theta_i)$ ($i = 1, ..., 439$), the primary problem was essentially to model the 439 parameters θ_i. In addition, there was the secondary problem indicated by Dr Briese, to find a satisfactory model for the data of Table 4.

We use subscript i to refer, as above, to all the progenies studied,

Table 3. *26 progenies initially allocated to Q-type group*

		Progeny k				
l	Locality l	k	m_k	f_k	n_k	Sex ratio m_k/n_k
4	Cotter River	1	0	51	51	0
5	Mildura/Merbein	2	0	26	26	0
8	Waikerie	3	18	40	58	0.310
15	Murray Bridge	4	0	21	21	0
	Murray Bridge	5	0	20	20	0
16	Tailem Bend	6	0	60	60	0
	Tailem Bend	7	0	51	51	0
	Tailem Bend	8	0	14	14	0
	Tailem Bend	9	1	54	55	0.018
	Tailem Bend	10	2	53	55	0.036
	Tailem Bend	11	7	47	54	0.130
18	Nhill	12	2	59	61	0.033
34	Tascott	13	0	33	33	0
36	Wangi Wangi	14	13	34	47	0.277
38	Muswellbrook	15	0	35	35	0
	Muswellbrook	16	0	48	48	0
	Muswellbrook	17	0	61	61	0
	Muswellbrook	18	0	42	42	0
	Muswellbrook	19	1	42	43	0.023
	Muswellbrook	20	1	29	30	0.033
45	Swansea	21	0	19	19	0
	Swansea	22	0	51	51	0
	Swansea	23	0	15	15	0
	Swansea	24	0	50	50	0
	Swansea	25	6	55	61	0.098
46	Bicheno	26	14	29	43	0.326

subscript j to refer to normal-type progenies and subscript k to refer to Q-type progenies. For the 413 observations $(m_j, f_j; n_j)$ $(j = 1, ..., 413)$ summarised in Table 2, the briefing from Dr Briese suggested the simple model of a common parameter θ near 0.5. Testing the null hypothesis

$$H_0: \theta_j = \theta \ (\forall j) \tag{1}$$

against the alternative H_1: unrelated θ_js, the difference between the maximised log likelihoods $\hat{L}_0$ and $\hat{L}_1$ under the two hypotheses is

$$\hat{L}_1 - \hat{L}_0 = (\sum_j m_j \log m_j + \sum_j f_j \log f_j - \sum_j n_j \log n_j)$$
$$- (M \log M + F \log F - N \log N) \tag{2}$$

Table 4. *Distribution of numbers of Q-type and normal-type females (initial allocation) collected at the 68 localities*

Q_l	N_l	$T_l = Q_l + N_l$	Number of localities
0	2	2	10
0	3	3	11
0	4	4	6 (7)
0	5	5	8 (9)
0	6	6	4
0	7	7	3
0	8	8	1
0	9	9	3
0	10	10	1
0	12	12	3
0	13	13	2
0	14	14	1
0	15	15	2
0	16	16	1
0	19	19	1
1	2	3	1
1	3	4	3 (2)
1	4	5	1 (0)
1	7	8	1
1	11	12	1
2	5	7	1
5	5	10	1
6	6	12	1
6	16	22	1
			68

Total number of Q-type females collected (provisional allocation): $7 \times 1 + 2 + 5 + 2 \times 6 = 26$. Figures in brackets give final allocation.

where $M = \Sigma m_j$, $F = \Sigma f_j$, $N = \Sigma n_j$. This comes to 198.66, so the *deviance* $2(\hat{L}_1 - \hat{L}_0)$ corresponding to the economisation of 412 parameters in using the simpler model is 397.3, which is clearly non-significant when tested as a value of χ^2_{412}.

Turning now to the 26 observations $(m_k, f_k; n_k)$ $(k = 1, ..., 26)$ in Table 3, it was obvious (and confirmed statistically below) that the parameters θ_k could not reasonably be taken to be *all* equal, although perhaps most of them could. But since from the entomological point of view only one category of non-normal condition was being posited, namely *the* Q-condition, a single model was called for which would fit all the non-normal sex ratios. A mixed-binomial model suggested itself, in which there was a population of θ_k-values corresponding to the population of adult Q-type females; the number of males m_k in a progeny of size n_k would then have a mixed-binomial distribution (over all Q-type progenies of this size), namely the average of the binomial distributions $B(n_k, \theta_k)$ with respect to the θ_k-distribution.

An obvious model to try for the mixing distribution, with support confined to $0 \leq \theta \leq 1$ as required, was a beta distribution, this being simple, tractable and flexible. Taking this distribution in the form

$$g(\theta) = \frac{\Gamma(\alpha+\beta)}{\Gamma(\alpha)\Gamma(\beta)} \, \theta^{\alpha-1}(1-\theta)^{\beta-1} \quad (0 \leq \theta \leq 1), \tag{3}$$

we get the new null hypothesis

$$H_0^*: P(m_k = m \,|\, n_k) = \frac{\Gamma(\alpha+\beta)}{\Gamma(\alpha)\Gamma(\beta)} \binom{n_k}{m} \frac{\Gamma(\alpha+m)\Gamma(\beta+f)}{\Gamma(\alpha+\beta+n_k)} \tag{4}$$

$(f = n_k - m; k = 1, ..., 26)$,

to be tested against the old alternative H_1: unconnected θ_k.

The log likelihood $L(\alpha, \beta)$ under H_0^* is, apart from a constant, of the form

$$L(\alpha, \beta) = \sum_k \{\log(\alpha+m_k-1)+\log(\alpha+m_k-2)+...+\log\alpha$$

$$+\log(\beta+f_k-1)+\log(\beta+f_k-2)+...+\log\beta$$

$$-\log(\alpha+\beta+n_k-1)-\log(\alpha+\beta+n_k-2)-...$$

$$-\log(\alpha+\beta)\}. \tag{5}$$

Thus the solution $\hat{\alpha}$, $\hat{\beta}$ of $\partial L/\partial\alpha = 0$, $\partial L/\partial\beta = 0$ was easily obtained by iterative calculations on sums of reciprocals, giving a maximised likelihood $L(\hat{\alpha}, \hat{\beta}) = \hat{L}_0^*$ under H_0^*. The value of this was -43.30, and the maximised likelihood under H_1 (unrelated θ_ks) was $\hat{L}_1 = -15.43$; so the deviance, to be tested as a value of χ^2_{24} for the fit of the two-parameter model (4) to the 26 sex ratios, was 55.7, a highly significant value ($P \approx 0.0025$).

In the light of this failure of the proposed model to fit the data satisfactorily, Table 3 was re-examined. Clearly while most of the observed sex ratios were at or near zero, three of them, listed in Table 5, stood out as unusually high in relation to the rest. These three values 18/58, 13/47 and 14/43 were in fact *upper outliers with respect to the model (4)* for the Q-type data. At the same time the initial classification of the data had placed them in Table 3 rather than Table 2 because they were unusually low in relation to $\frac{1}{2}$; in other words, they had been declared to be *lower outliers with respect to the model (1)*. Could a home now be found for them, if not under one roof then under the other?

Note the difference from the usual problem of outlier identification, in which we have an outlying value (or set of outlying values) which are extreme with respect to some main population; we wish to judge by a discordancy test whether or not it is statistically reasonable to regard the outlier (or outliers) as belonging, though extreme, to the main population. If found not reasonable then the outlier is judged *discordant* and requires a separate probability model from the main population. Here we have a different problem. Our outlying values lie *between two main populations* and are extreme with respect to *both*; we wish to *allocate* each outlier to one or other of the populations and thus avoid setting up separate models for them. Is this statistically reasonable, and if so how should the allocation be done?

On taking a further look at Table 5, it appeared that the values of m_k, while much lower than the respective values of $\frac{1}{2}n_k$, were not after all very implausible as observations from binomial distributions $B(n_k, \theta)$ with $\theta \approx \frac{1}{2}$ in a large data set of size 416 say. For the Waikerie progeny, $P(m_k \leq 18) = 0.0029$ assuming a $B(58, \frac{1}{2})$ distribution; the corresponding lower tail probabilities for the Wangi Wangi and Bicheno progenies are 0.0018 and 0.0164. An approximate discordancy test can be carried out for the three outliers *en bloc*, using the test statistic

$$T_{N\mu\sigma3} = [3\mu - x_{(1)} - x_{(2)} - x_{(3)}]/\sigma \tag{6}$$

given in Barnett and Lewis (1984, p. 189; or 1978, p. 112). Here the outliers are $x_{(1)} = 0.277$, $x_{(2)} = 0.310$, $x_{(3)} = 0.326$, and they are to be tested for

Table 5. *Three progenies requiring possible re-allocation*

k (Table 3)	Locality	m_k	f_k	n_k	Sex ratio
3	Waikerie	18	40	58	0.310
14	Wangi Wangi	13	34	47	0.277
26	Bicheno	14	29	43	0.326

discordancy in relation to an assumed normal distribution $N(\mu, \sigma^2)$ with known mean μ (here $\mu = 0.5$) and known variance σ^2. In fact, the three binomial proportions have different variances on the null hypothesis, namely $(0.5)^2/47$, $(0.5)^2/58$ and $(0.5)^2/43$, so giving σ^2 the smallest of these values ensures a conservative test. The value of the test statistic (6) is then

$$(1.5-0.913)\sqrt{58}/(0.5) = 8.94.$$

This is to be judged in relation to the tabulated 5% and 1% points given in Barnett and Lewis (1984, p. 385; or 1978, p. 306). Our sample size is $n = 416$, but the entries in the table stop at $n = 100$:

	5% point	1% point
$n = 40$	7.15	7.90
$n = 50$	7.34	8.17
$n = 100$	8.11	8.83

Obviously 8.94 is non-significant for $n = 416$, so the three progenies of Table 5 are consistent with the main normal-type data set of Table 2.

At the time, I revised the calculation of deviance (2) to include the three progenies in Table 5; the number of assumed normal-type progenies went up from 413 to 416, and I found $2(\hat{L}_1 - \hat{L}_0) = 422.53$ which again is non-significant as a value of χ^2_{415}. For these 416 progenies $M = 10613$, $N = 20976$, and a common value θ could be assumed for the underlying sex ratios θ_j, with a point estimate $\hat{\theta} = 10613/20976 = 0.506$ and approximate 95% confidence limits

$$0.506 \pm 1.96\sqrt{(0.506 \times 0.494/20976)},$$

i.e. $0.499, 0.513.$ (7)

Were there any further progenies besides the above three that could be reclassified from Q-type to normal-type? No. The least unlikely candidate from Table 3 was $m_k = 7, f_k = 47$ ($k = 11$), with sex ratio $7/54 = 0.130$. This can be tested for discordancy, to a good approximation, as a lower outlier in a sample of size 417 from a normal distribution with known mean $\mu = 0.5$ and known variance σ^2, choosing for σ^2 the value $(0.5)^2/54$ (note that the mean of the 417 sample sizes n_j is 50.4, so the test is on the conservative side). The test statistic $T_{N\mu\sigma1}$ given in Barnett and Lewis (1984, p. 188; or 1978, p. 111) has value -5.24, highly significant ($P \approx 0.00004$).

The transfer of these three progenies to the normal-type group removed the three entries $k = 3$, $k = 14$ and $k = 26$ from Table 3 and reduced the number of proposed Q-type progenies from 26 to 23. Repeating the calculations based on (4) and (5) for the reduced set, maximum likelihood estimates of α and β in (4) were obtained as $\hat{\alpha} = 0.331, \hat{\beta} = 18.15$, giving

$\hat{L}_0^* = L(\hat{\alpha}, \hat{\beta}) = -88.21$, while $\hat{L}_1$ corresponding to unrelated θ_ks was -71.96. The deviance, to be tested as a value of χ^2_{21} for the fit of model (4) to the 23 sex ratios, was thus 32.49. This value is rather large ($P = 5.2\%$) but not so large as to discredit the mixed-binomial model (4).

It seemed obvious to me, in the light of the above analysis, that the three 'borderline' progenies in Table 5 were clear candidates for exclusion from the Q-type set and inclusion in the normal-type set, and I reported to the entomologists in the following terms:

> ...statistical analysis...makes it clear that the three observations, $(m, n) = (14, 43)$, $(18, 58)$ and $(13, 47)$, must be allocated to the normal group, making its size 416 and reducing the number of Q-type females in the survey to twenty-three. ...The allocation of all the observations to normal or Q-type is...clear cut and unambiguous in this analysis.

I was wrong! In a further part of Dr Geier and Dr Briese's experimental programme, observations were made of the sex ratios in 'second-generation' progenies produced by female moths belonging to some of the 'first-generation' progenies ($k = 9, 10, 11, 12, 25, 26$) in Table 3. One such daughter moth in the Bicheno progeny $m = 14, f = 29, n = 43$ produced progeny $m = 1, f = 14, n = 15$, establishing that the original field-collected mother was of Q-type, and not of normal type! So it emerges, at the end of the day, that the allocation of the three progenies in Table 5 should be two (Waikerie and Wangi Wangi) to normal and one (Bicheno) to Q-type.

We end up with 415 normal-type progenies out of the 439; the value of $\hat{\theta}$ and the 95% confidence limits for θ remain as in (7). What about the fit of model (4) to the enlarged set of 24 Q-type sex ratios, a model which I had previously been sure could not be extended to cover more than the extreme 23? Including now the Bicheno sex ratio 14/43, the revised estimates of the parameters were

$$\hat{\alpha} = 0.209, \quad \hat{\beta} = 6.65.$$

The information matrix was readily estimated from the second derivatives of (5), giving the estimated standard errors (SE) and correlation as

$$\mathrm{SE}(\hat{\alpha}) = 0.105, \quad \mathrm{SE}(\hat{\beta}) = 4.28, \quad \hat{\rho}(\hat{\alpha}, \hat{\beta}) = 0.709$$

The deviance $2(\hat{L}_1 - \hat{L}_0^*)$ for the fit of model (4) to the 24 sex ratios came out to 42.24. Testing this as a value of χ^2_{22} we get a P-value of 0.0058 or about 1 in 170. Under the guidance of the overriding entomological evidence, we accept that an event with this small (but not incredibly small) outside chance has occurred.

4   Modelling the Q-type ratios

One question remained, the 'secondary problem' raised by Dr Briese at the end of his letter: could a satisfactory model be found for the proportions of Q-type adults in the samples collected at the 68 different localities? The data are in Table 4; the four revised frequencies given there in brackets correspond to the reclassification from Q to N of the two 'borderline' females collected at Waikerie and Wangi Wangi.

On examination, one noted a remarkable similarity between the distribution over localities of the proportions Q_l/T_l of Q-type females (Table 4) and the distribution over Q-type females of the proportions m_k/n_k of male progeny (Table 3). Using the figures for the final allocation to Q-type and normal groups, characteristics of the two samples are as follows:

	The 68 Q-type ratios Q_l/T_l	The 24 sex ratios m_k/n_k
Mean	0.038	0.029
Standard deviation	0.110	0.071
Skewness coefficient	2.97	3.26
Peakedness coefficient	11.0	13.3

For whatever reason the distributions appear to be remarkably similar in shape. This suggested it would be reasonable to model the two data sets in a similar way. Just as a binomial model $B(n_k, \theta_k)$ can reasonably be assumed for m_k, so a binomial model $B(T_l, \phi_l)$ can reasonably be assumed for Q_l, where the ϕ_ls $(l = 1, ..., 68)$ are parameters associated with the respective localities l. It was then natural to attempt to fit the data (Q_l, T_l) by a mixed-binomial model of form (4), based on a mixing distribution of form (3) for the ϕ_ls. This proved to be a very satisfactory fit; curiously enough, the data (Q_l, T_l) and the data (m_k, n_k) could reasonably be fitted by the same distribution.

Acknowledgements

I am most grateful to Dr P. W. Geier and Dr D. T. Briese for giving me the opportunity to work on this problem, for valuable discussions, and for generously giving permission for their data and correspondence to be quoted here.

References

Barnett, V. and Lewis, T. (1978, 1984) *Outliers in Statistical Data*, 1st and 2nd editions, Wiley.

Geier, P. W. and Briese, D. T. (1977) Predominantly female progeny in the light-brown apple moth. *Search*, **8**, 83–5.

Geier, P. W. and Oswald, L. T. (1977) The light-brown apple moth, *Epiphyas postvittana* (Walker). 1. Effects associated with contamination by a nuclear polyhedrosis virus on the demographic performance of a laboratory strain. *Aust.J.Ecol.*, **2**, 9–29.

Geier, P. W. and Springett, B. P. (1976) Population characteristics of Australian leafrollers infesting orchards (*Epiphyas* spp., Lepidoptera). *Aust.J.Ecol.*, **1**, 129–44.

Geier, P. W., Briese, D. T. and Lewis, T. (1978) The light-brown apple moth, *Epiphyas postvittana* (Walker). 2. Uneven sex ratios and a condition contributing to them in the field. *Aust.J.Ecol.* **3**, 467–88.

10

Collaboration between university and industry

B. J. T. MORGAN, P. M. NORTH AND S. E. PACK

1   Introduction

In this paper we describe two of the several different ways in which statisticians in a university can act as consultants for industry. In both cases the consulting is effectively carried out by more than one statistician, but there the similarity ends. We hope that these examples will provide a flavour of the activities of an applied statistics department and an applied statistics research unit working together within a university. A number of problems are considered, including: the analysis of dominant lethal assay data; the analysis of quantal assay data incorporating time to response; the analysis of pain data relating to episiotomy; the analysis of aggression in mentally handicapped patients.

2   The analysis of dominant lethal assay data

Tables 1 and 2 present two sets of data from the paper by Haseman and Soares (1976). In each case, for over 500 litters of mice, the number of dead fetuses was recorded. Tables 1 and 2 are control groups from dominant lethal assays (taken from Haseman and Soares, 1976). In this experiment a drug's ability to cause damage to reproductive genetic material, sufficient to kill the fertilised egg or developing embryo, is tested by dosing a male mouse (typically) and mating it to one or more females. A significant increase in fetal deaths is indicative of a mutagenic effect.

As in many areas of statistics, typically two questions arise:

(1) Can we describe such sets of data in a relatively simple manner?

(2) How might we make comparisons between such data sets?

In the simplest response to (2), t-tests may be used to make the comparison. Following a Freeman–Tukey binomial transformation, a t-test results in an approximately standard normal test statistic of 2.04, significant at the 5% level. Alternatively, a t-test based on Kleinman's

Table 1. *Population no. 1 of Haseman and Soares: observed frequency distribution of fetal death and fitted frequencies under the beta-binomial*

Litter size (n)	Number of dead fetuses (x)													
	0	1	2	3	4	5	6	7	8	9	10	11	12	13
1	2													
	1.8	0.2												
2	2													
	1.7	0.3												
3	3													
	2.3	0.6	0.1											
4	5	1	1											
	5.0	1.6	0.4	0.1										
5	2	2												
	2.6	1.0	0.3	0.1										
6	2	2												
	2.5	1.0	0.4	0.1										
7	2	2	2	1										
	4.1	1.9	0.7	0.2	0.1									
8	6	1		1	1									
	4.9	2.5	1.1	0.4	0.1									
9	2	3	1											
	3.1	1.7	0.8	0.3	0.1									
10	2	4	2	2										
	4.9	2.8	1.4	0.6	0.2	0.1								
11	19	11	3	3										
	16.6	10.8	5.2	2.5	1.1	0.4	0.2							
12	33	24	11	5	4	4							1	
	35.9	22.6	12.4	6.2	2.9	1.2	0.5	0.2	0.1					
13	39	27	12	6	5	2			1					
	38.3	25.1	14.4	7.6	3.7	1.7	0.7	0.3	0.1					
14	34	30	14	6	6			1						
	36.2	24.5	14.6	8.0	4.2	2.0	0.9	0.4	0.2	0.1				
15	38	22	18	4	2	1								
	32.3	22.6	13.9	7.9	4.3	2.2	1.1	0.5	0.2	0.1				
16	13	16	14	4	3	1								
	18.5	13.3	8.4	5.0	2.8	1.5	0.8	0.4	0.2	0.1				
17	8	4	3	3	2	1		1						
	7.7	5.6	3.7	2.2	1.3	0.7	0.4	0.2	0.1					
18		4	2	1										
	2.3	1.8	1.2	0.7	0.4	0.3	0.1	0.1						
19	2	1												
	1.0	0.7	0.5	0.3	0.2	0.1	0.1							
20													1	
	0.3	0.2	0.2	0.1	0.1									

weighted estimator (Kleinman, 1973) produces a corresponding test statistic of 2.61, significant at the 1% level.

A less pragmatic approach is to model each data set, in response to (1), and then make comparisons using the fitted models. Within any one litter one might suppose that deaths have a binomial distribution with probability p, say, of death for each fetus. A model for which p does not vary over

Table 2. *Population no. 3 of Haseman and Soares: observed frequency distribution of fetal death and fitted frequencies under the beta-binomial*

Litter size (n)	Number of dead fetuses (x)									
	0	1	2	3	4	5	6	7	8	9
1	7									
	6.5	0.5								
2	7									
	6.0	0.9	0.1							
3	6									
	4.8	1.0	0.2							
4	5	2	1							
	6.1	1.5	0.3	0.1						
5	8	2	1		1	1				
	9.3	2.8	0.7	0.2						
6	8									
	5.4	1.8	0.6	0.2						
7	4	4	2	1						
	7.1	2.6	0.9	0.3	0.1					
8	7	7	1							
	9.2	3.7	1.4	0.5	0.2					
9	8	9	7	1	1					
	15.3	6.5	2.7	1.0	0.4	0.1				
10	22	17	2		1			1	1	
	24.8	11.0	4.9	2.1	0.8	0.3	0.1			
11	30	18	9	1	2		1		1	
	33.5	15.7	7.3	3.3	1.4	0.6	0.2	0.1		
12	54	27	12	2	1		2			
	50.9	24.8	12.2	5.8	2.6	1.1	0.4	0.2	0.1	
13	46	30	8	4	1	1		1		
	45.5	23.0	11.7	5.9	2.8	1.3	0.5	0.2	0.1	
14	43	21	13	3	1			1		1
	40.0	20.9	11.1	5.7	2.9	1.4	0.6	0.3	0.1	
15	22	22	5	2	1					
	24.2	13.0	7.1	3.8	2.0	1.0	0.5	0.2	0.1	
16	6	6	3		1	1				
	7.6	4.2	2.4	1.3	0.7	0.4	0.2	0.1		
17										
18	3		2	1						
	2.5	1.5	0.9	0.5	0.3	0.2	0.1			

litters is usually regarded as far too simple for data of this kind, and in one approach, considered by Williams (1975), p is given a beta distribution over litters.

Under the beta-binomial model, the probability of x dead fetuses in a litter of n mice is given by

$$\Pr(x) = \binom{n}{x} \frac{B(x+\alpha, n-x+\beta)}{B(\alpha, \beta)}, \quad \text{for } 0 \leqslant x \leqslant n,$$

where α, β are the parameters of the beta distribution assumed for the binomial probability of response. If we adopt a more stable parameterisation in terms of $\mu = \alpha/(\alpha+\beta)$, and $\theta = (\alpha+\beta)^{-1}$ then the beta-binomial mean and variance are, respectively, $n\mu$, and $n\mu(1-\mu)(1+\phi(n-1))$, with the parameter $\phi = \theta/(1+\theta)$ representing the necessary variance-inflation, relative to the binomial distribution, with results as $\theta \to 0$.

The fitted values are shown in Tables 1 and 2, and clearly provide a good qualitative description of the data. We obtain the parameter estimates given below together with asymptotic measures of error (μ_i, θ_i denote the parameter values for group i, and L_i denotes the maximised log-likelihood, $i = 1, 2$).

$$\mu_1 = 0.0901 \ (0.0045), \quad \mu_2 = 0.0739 \ (0.0042)$$
$$\theta_1 = 0.0730 \ (0.0107), \quad \theta_2 = 0.0813 \ (0.0123)$$

For each set of data we see that the binomial model, resulting as $\theta \to 0$, would give a poor fit, as standard confidence intervals for θ_i do not contain $\theta_i = 0$, $i = 1, 2$. See also Table 3; justification of a likelihood-ratio test is given in Prentice (1986).

The maximised log-likelihood values are $L_1 = -777.79$, and $L_2 = -701.33$. As a quantitative test of goodness-of-fit, we simulate 99 additional sets of data, separately from each group, simulating from the fitted beta-binomial models above, and for each data set computing the maximum log-likelihood values. We find that L_1 is ranked 33rd largest out of the sample of 100 values of L_1, with values ranging from -734.22 to -815.25, and L_2 is ranked 46th largest in the corresponding sample of 100 values of L_2, with values ranging from -634.42 to -746.55. In each case therefore these values confirm the earlier impression of a satisfactory fit to the data.

In using the models for comparing the two data sets we may be interested in testing hypotheses of the kind:

Case 1 H_0: $\mu_1 = \mu_2, \theta_1 = \theta_2$ versus H_1: $\mu_1 \neq \mu_2, \theta_1 \neq \theta_2$

Case 2 H_0: $\mu_1 = \mu_2, \theta_1 \neq \theta_2$ versus H_1: $\mu_1 \neq \mu_2, \theta_1 \neq \theta_2$.

For Case 1, the likelihood-ratio test statistic is 8.25, and the Wald test statistic is 8.26, results significant at the 2.5% level. For Case 2, the likelihood-ratio test statistic is 6.16, and the Wald test statistics is 6.25, both results again significant at the 2.5% level. (For details of the Wald test see, for example, Silvey (1975, p. 115).)

These three different approaches to comparing the data of Tables 1 and 2 thus all result in conclusions of a significant difference, but with varying strengths of significance. While the beta-binomial model provides an adequate description of the data, one might also wonder whether other models might provide a better fit. For example, Altham (1978) and Kupper and Haseman (1978) propose a correlated binomial distribution, and Paul (1984) and Pack (1986a) consider the beta-correlated binomial distribution.

Fitting a variety of models to the data of Tables 1 and 2 produces the maximum log-likelihood values of Table 3. These values emphasise the earlier conclusion that the standard binomial distribution is unsuitable for these data. There is a suggestion that the beta-binomial is better than the correlated binomial, but that the beta-binomial fit for the second data set could be improved.

Obvious questions arise from the above analyses: need a statistician become involved in fitting fairly complex probability models to such data, and, if so, at what stage should the model development end? Might much simpler *t*-tests be adequate for comparisons? A review of the literature reveals a large amount to recent research activity in this area. Smith (1983) provides an algorithm for fitting the beta-binomial, and an approximate approach using GLIM is described by Brooks (1984). Pregibon (1982) considers robust analysis. Vuataz and Sotek (1978) and Paul (1982) suggest that in general the beta-binomial fits data better than the correlated binomial model, yet James and Smith (1982) encounter problems fitting

Table 3. *Maximised log-likelihood values from fitting a variety of models to the data of Tables 1 and 2*

	Model				
	Binomial	Correlated binomial	Beta-binomial	Beta-correlated binomial	Mixture of two binomials
Table 1	−842.62	−801.69	−777.79	−776.38	−782.45
Table 2	−745.06	−732.19	−701.33	−695.94	−686.23

the beta-binomial model, finding it lacks robustness with respect to outlying values. More generally, from papers such as: Vuataz and Sotek (1978), Gladen (1979), Haseman and Soares (1976), Haseman and Kupper (1979) and Shirley and Hickling (1981), there is conflicting evidence regarding power and type I error rates.

More typical than the data sets of Tables 1 and 2 are experiments resulting in 20–40 litters. From investigation of real and simulated data sets of this kind, Pack (1985, 1986a, b) has been able to conclude that for much data arising from the dominant lethal assay, the most useful models are likely to be the beta-binomial and a mixture of binomials. For comparative experiments the beta-binomial is likely to be preferred, and overall it has been found in this case that likelihood-ratio tests are at least as powerful as the simpler t-tests mentioned above, and in certain situations they can be significantly more powerful.

The time necessary to review the literature, consider the problems and conduct the experiments before such conclusions can be reached is unlikely to be available to the average consultant. In this case the work was done by the Scientific Computing and Statistics Department of the Wellcome Research Laboratories at Beckenham, collaborating with the University of Kent through a linked Science and Engineering Research Council studentship (CASE award), to allow a research student, Simon Pack, to work for a PhD in statistics through consideration of these problems. A particular feature of the Wellcome data was the repeat mating of male rats with different females over time. This led to a broadening of the research work, and to a consideration of other toxicology experiments in which time plays an important role. We shall consider one such example in the next section.

3   Analysis of quantal assay data with time to response

Table 4 presents examples of the data to be considered here. These data were analysed by Diggle and Gratton (1984), who used Monte Carlo inference to fit the implicit stochastic model of Puri and Senturia (1972). An alternative approach has been suggested recently by Carter and Hubert (1984), and Jarrett (1984) has investigated a 'MICE' index which is sometimes used to summarise such data. An early analysis of similar data is provided by Boyce and Williams (1967).

The data can be viewed as constrained and censored survival data – see, for example, Aranda-Ordaz (1983) and Wolynetz (1979) – and it is a survival analysis approach that we developed here.

Let us suppose that we have D dose levels, $\{d_i\}$, and also that observations are taken at times $\{t_j\}$, with n_{ij} individuals responding in the interval, (t_{j-1}, t_j), for $1 \leqslant i \leqslant D$.

If the probability that an individual beetle given dose d_i dies in (t_{j-1}, t_j) is written as p_{ij}, then we model p_{ij} by:

$$p_{ij} = F(t_j, d_i) - F(t_{j-1}, d_i),$$

with

$$F(t_j, d_i) = \frac{a_i}{1 + e^{-\eta_{ij}}}, \tag{1}$$

where a_i is a function of the dose d_i, but η_{ij} also incorporates time to response.

Experimentation with fitting different models resulted in the following model to describe the data for males and for females:

$$F(t_j, d_i)^{-1} = (1 + e^{-(\alpha_1 + \alpha_2 \log d_i)}) (1 + t_j^{-\phi}\beta).$$

For the males, the maximum-likelihood parameter estimates, with estimated asymptotic standard errors, are

$$\hat{\alpha}_1 = 4.63 \ (0.46),$$
$$\hat{\alpha}_2 = 3.37 \ (0.33),$$
$$\hat{\phi} = 2.70 \ (0.14),$$
$$\hat{\beta} = 14.28 \ (2.38),$$

Table 4. *Hewlett's flour beetle data*

Dose (mg/cm²):	0.20		0.32		0.50		0.80	
Sex:	M	F	M	F	M	F	M	F
Group size:	144	152	69	81	54	44	50	47
Time (days)								
1	3	0	7	1	5	0	4	2
2	14	2	17	6	13	4	14	9
3	24	6	28	17	24	10	22	24
4	31	14	44	27	39	16	36	33
5	35	23	47	32	43	19	44	36
6	38	26	49	33	45	20	46	40
7	40	26	50	33	46	21	47	41
8	41	26	50	34	47	25	47	42
9	41	26	50	34	47	25	47	42
10	41	26	50	34	47	25	48	43
11	41	26	50	34	47	25	48	43
12	42	26	50	34	47	26	48	43
13	43	26	50	34	47	27	48	43

with estimated asymptotic correlation matrix

$\hat{\alpha}_1$	0.976		
$\hat{\alpha}_2$	-0.171	-0.152	
$\hat{\phi}$	-0.105	-0.093	0.834
	$\hat{\alpha}_2$	$\hat{\phi}$	$\hat{\beta}$

For the females, the maximum-likelihood parameter estimates, with estimated asymptotic standard errors, are

$\hat{\alpha}_1 = 2.61$ (0.27),

$\hat{\alpha}_2 = 2.61$ (0.22),

$\hat{\phi} = 3.48$ (0.20),

$\hat{\beta} = 57.05$ (14.83)

with estimated asymptotic correlation matrix,

$\hat{\alpha}_1$	0.937		
$\hat{\alpha}_2$	-0.047	-0.031	
$\hat{\phi}$	-0.035	-0.023	0.906
	$\hat{\alpha}_2$	$\hat{\phi}$	$\hat{\beta}$

Fitted values are given in Table 5. The agreement appears to be good, especially as the model only contains four parameters. The residual deviances and corresponding degrees of freedom are: for males 48.39(48),

Table 5. *Hewlett's flour beetle data: fitted values*

Dose (mg/cm²):	0.20		0.32		0.50		0.80	
Sex:	M	F	M	F	M	F	M	F
Group size:	144	152	69	81	54	44	50	47
Time (days)								
1	3.0	0.4	3.1	0.6	3.2	0.5	3.2	0.7
2	14.1	4.2	14.9	5.4	15.4	5.0	15.3	6.8
3	26.1	11.5	27.5	14.8	28.3	13.5	28.3	18.4
4	33.8	17.7	35.6	22.8	36.7	20.8	36.6	28.4
5	38.1	21.3	40.2	27.4	41.4	25.0	41.4	34.3
6	40.6	23.2	42.8	29.9	44.1	27.3	44.0	37.3
7	42.1	24.2	44.3	31.2	45.7	28.5	45.6	39.0
8	43.0	24.8	45.2	31.9	46.7	29.2	46.6	39.9
9	43.5	25.1	45.9	32.4	47.3	29.6	47.2	40.4
10	43.9	25.3	46.3	32.6	47.7	29.8	47.6	40.7
11	44.2	25.5	46.6	32.8	48.0	30.0	47.9	41.0
12	44.4	25.6	46.8	32.9	48.2	30.1	48.2	41.1
13	44.6	25.6	46.9	33.0	48.4	30.1	48.3	41.2

and for females 49.94(48), again indicative of a good fit. The chi-square approximation for the deviance was checked using simulation, and found to be satisfactory.

Fitting the same model to both male and female data sets produced a deviance of 139.91 on 100 degrees of freedom (d.f.), and so a likelihood-ratio test statistic of the hypothesis that $(\alpha_1, \alpha_2, \phi, \beta)$ does not vary with sex is 41.58, extremely significant when referred to chi-square tables on four degrees of freedom.

One way of investigating this difference further is as follows: we can assume that only (α_1, α_2) does not vary with sex. The residual deviance is then 128.38 on 98 degrees of freedom, highly indicative of a significant effect of sex on (α_1, α_2), suggesting in fact that males are more susceptible than females. Conversely, if we assume that only (ϕ, β) does not vary with sex the residual deviance is then 110.70 on 98 degrees of freedom, suggesting that the distribution of time-to-death differs between sexes, males having a smaller mean time-to-death than females. Both of these conclusions were reached also by Diggle and Gratton (1984), after a much more complicated analysis.

In the example considered here, a fraction of the flour beetles survived at each dose. In other examples (see, for example, Kooijman, 1981; Carter and Hubert, 1984) for which that does not appear to be the case, simpler models of the form

$$\text{logit } F(t_j, d_i) = \gamma_1 + \gamma_2 \log t_j + \gamma_3 \log d_i$$

have been fitted with success to describe the data. These are proportional odds models (see McCullagh, 1980; Bennett, 1983). An alternative approach which has also been useful is to fit proportional hazard models (see Cox, 1972), and work is continuing on the relative merits of these two different types of model for this kind of data.

Here, as in the example of the last section, probability models are constructed to describe sets of data in as parsimonious a way as possible, and then provide a suitable framework for comparisons.

4   Aspects of CASE awards

The advantages and disadvantages of CASE studentships have been detailed by Jones, Morgan and Wetherill (1983), and the Wellcome CASE award is no exception. The Wellcome statistician, David Smith, has provided constant and invaluable input into the research programme. However, active dominant lethal toxicology work ceased at Beckenham virtually as the CASE studentship began. This eventuality was not anticipated during the lengthy period of time in which the case for the award was prepared, the grant application made and approved, and a student secured. A wealth of historical data provided an excellent start to the project, but the anticipated interaction with active toxicologists was lacking. Fortunately the topic of Section 3, which did not arise directly from the Wellcome data, led on to consider experiments in which knocked-down insects could recover, and by chance an excellent supply of data from such experiments was available from Wellcome (but now at Berkhamsted). Work is currently underway on modelling these data, which seem to require implicit models, in the terminology of Diggle and Gratton (1984). CASE studentships frequently take the supervisor, as well as the student, into uncharted waters. A problem with research in an unfamiliar area is that one may not be as aware as one should be of recent relevant work. Thus the paper of Farewell (1982), in which he independently proposes the model (1), was only discovered, after the work had been done, as a footnote to the discussion of Kalbfleish, Krewksi and Van Ryzin (1983).

The Wellcome CASE award grew out of contact with the Applied Statistics Research Unit (ASRU) at Kent, which in turn developed as a result of our extensive industrial links, quite often through CASE awards (see Jones, Morgan and Wetherill, 1983). ASRU now provides another way in which the university may collaborate with industry. The luxury of three months, let alone three years, for consultancy work done by ASRU is simply not possible, as the examples of the next two sections demonstrate.

5   ASRU consultancy on clinical trials

ASRU is a self-financing group which therefore needs to attract regular contract work from industry, government, research stations, etc. for its continuing existence. It has to be flexible enough to be able to meet work demands as they currently exist at any given time, in whatever area of statistics. One area that has always featured prominently in ASRU's work is that of clinical trial analysis, a subject which is also discussed elsewhere in this volume.

For such work there is certainly not time available to carry out extensive literature searches, try out increasingly complex modelling approaches and, perhaps, develop new methodology, as there would be in, for example, CASE projects such as the one described in Sections 2 and 3. The client company is now the paymaster and typically requires a rather specific piece of work to be carried out. The request rarely extends beyond the analysis of the data arising out of the trial in question, with no provision for possibly interesting and desirable follow-up work. As the earlier sections indicate, however, other means exist for organising such research. Occasionally, a client's request may be very specific, in which case it is conceivable that it might contravene advice that would have been given by ASRU. As an agent contracted to do the specified work, ASRU would then still give its advice, but carry out the work it is contracted to do unless this is really diametrically opposed to its own judgement! This is one feature of consultancy that is not likely to occur in any other type of university–industry collaboration. Most importantly, client companies often require the work to be carried out to very tight deadlines. Since requests may arise at any time of the year, it is only by having a group such as ASRU that a university is then in a position to meet such demands. Reports may have to be produced in standard style, and contain familiar methodology, or novel methods comprehensively referenced or described, often for eventual submission to drug regulatory authorities. While this kind of observation may come as second nature to an industrial statistician, such concepts may be quite new to a statistician with a more exclusively academic background. Reports usually have to be written so that they can be understood by non-statisticians, for example clinicians, as well as by statisticians. ASRU work provides its statisticians with regular opportunities to interact with non-statisticians (but this is likely to be true of much statistical consultancy anyway). However, on the other hand, ASRU consultancy is also sometimes carried out with a client company's statistician as the direct contact, leaving the ASRU statistician at least one person removed from the experimenter: this is less likely to be true in more standard consultancy work. A further feature which is particularly pertinent to ASRU's work is the need to respect absolutely the confidentiality of its arrangements with clients – who at any one time may include a number of competing companies within, for example, the pharmaceutical industry.

The data analysis tasks undertaken in the clinical trials area by ASRU on behalf of clients are often fairly straightforward in nature, though sometimes slightly non-standard features exist, or points of interest arise to be followed up. For example, in a recent study questioning the necessity of routine episiotomy for mothers at delivery (Harrison *et al.*, 1984), the pain score data listed in Table 6 were obtained for the group of patients undergoing episiotomy and for the group sustaining a second-degree tear. A comparison between the distributions of pain scores in the two groups can be made at each measurement time by constructing the corresponding contingency tables, in which patients are classified by 'treatment' group and severity of pain. (Note that this itself already introduces the problems of multiple testing and non-independence of tests, which bedevil a lot of clinical trials analysis. Further comment on repeated measurement analysis is made below.) A standard analysis of one of the resulting contingency tables would consist of computing the Pearson's chi-squared statistic for the data, and assessing that for significance. For data such as those given in Table 6, however, such an analysis would be wasteful of the information inherent in the ordered nature of the pain score data. Methods for the analysis of ordered contingency tables have already been well documented (see, for example, Everitt, 1977, pp. 51 and 100 and the sections following them) but, more recently, the models of McCullagh (1980), and the program for fitting them, PLUM, provide a useful approach to such data. An underlying continuous distribution of pain severity is postulated for each group of patients, and a test (based on analysis of deviance) for difference in location of the distributions is carried out. We would expect such a test to have more power in detecting group differences than the standard Pearson's chi-square approach, though in the case of the data of Table 6 significant differences (at the 5% level) are still not detected. Simple inspection of the data reveals that in this example the result is not at all surprising.

The analysis of repeated measurements is an area which is frequently encountered in ASRU clinical trials work. Figure 1 illustrates a common situation. It shows the patient group mean profiles of aggression scores in a trial lasting 12 weeks (plus 4 weeks run-in) and in which observations were recorded daily, later to be summarised by weekly periods. This trial was a multicentre parallel-group study, with five centres taking part, and

Table 6. *Severity of pain recorded over first four days after delivery in 77 patients (figures are numbers of patients)*

	Day 1		Day 2		Day 3		Day 4	
Pain	a.m.	p.m.	a.m.	p.m.	a.m.	p.m.	a.m.	p.m.
Patients undergoing episiotomy (n = 40)								
None	3	2	6	12	16	17	24	26
Mild	18	17	24	14	15	14	11	8
Moderate	10	14	8	14	7	7	3	4
Severe	8	6	2	—	2	2	1	2
Very severe	1	1	—	—	—	—	—	—
Patients sustaining second-degree tear (n = 37)								
None	7	5	5	6	13	16	17	22
Mild	15	15	22	20	15	12	16	9
Moderate	11	13	7	8	8	7	4	6
Severe	2	2	3	3	1	2	—	—
Very severe	2	2	—	—	—	—	—	—

was designed to assess the efficacy of a treatment, with placebo control, in the control of aggression in mentally handicapped patients. Acts of aggression were scored by hospital staff on a scale of 1 to 5, according to the following key: 1, well behaved; 2, mood uncertain; 3, overt regression or attempted aggression; 4, additional medication required to control patient; 5, seclusion required. The clinician may ask the ill-defined question of whether the profiles are significantly different. If so, the clinician might be interested to know when they first became significantly different and, in some instances, when the profiles first start coming together again. Although such questions may appear to the clinician to be very straightforward, they are not simple to handle statistically. Work on answering such questions is currently being undertaken at Kent.

Repeated measurement analysis procedures are readily available at present; an example is the BMDP 2V procedure (Dixon, 1981). This provides a test for the difference between the global means, and also for linear, quadratic, or high-order component of the time–treatment interaction. The Kent work (Kenward, in preparation) involves a step-by-step analysis through time, the successive tests that are carried out being conditioned on the previous ones. Forward and backward stepping is envisaged. With such follow-up research work, a method of funding, outside the normal ASRU contractual arrangement with a client, usually

Fig. 1. Group means and standard errors of mean weekly aggression score

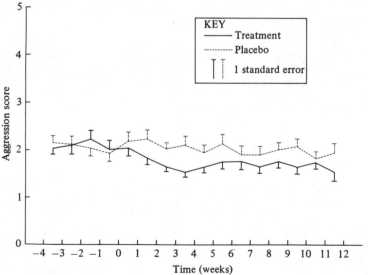

needs to be found. The work on repeated measurement was carried out by a member of the statistics department, illustrating again how the department and the Unit can work together to mutual benefit.

Many clinical trials have patient withdrawals. If there are many of these and, especially, if they may be treatment related, great care is needed in carrying out the data analysis. Such an example arose in a study recently handled by ASRU. The study, concerning patients suffering from rheumatoid arthritis, involved 20 centres in each of which 16 patients were to be recruited and allocated to one of three treatment groups (two active treatments, one placebo). In fact, only 224 patients from 15 centres entered the trial and, of these, only 92 completed it. In this particular example the early stopping of the trial because of the decision to stop development of the drug under study (a decision which was made independently of the trial) led to a large number of 'withdrawals'. However, in addition, the data for some visits had to be excluded from analysis because of the non-compliance of patients with the visit times laid down. Table 7 shows the number of patients in the trial at each visit whose data could be analysed.

Gould (1980) suggested an approach for handling data from trials where there are a substantial number of withdrawals. He suggests that before this, even as recently as 1976 (Peto *et al.*, 1977) there appeared to be no satisfactory method described in the literature. His own approach consists in ordering the reasons for withdrawal, for example, from adverse reaction to treatment to complete recovery, and assigning scores with a corresponding ordering to be used as imputed values in the analysis. Gould's approach, then, provides a useful start in the search for appropriate methodology for handling the treatment-related withdrawal problem. There is little doubt, however, that it would repay further research. This

Table 7. *Numbers of patients at each visit of a rheumatoid arthritis study with data appropriate for analysis: visits ranged over one year*

Trial visit number	Control group	Treatment group A	Treatment group B
1	56	55	112
2	53	53	107
3	51	53	98
4	47	49	90
5	42	40	80
6	25	30	67
7	20	20	51

again illustrates how ASRU consultancy can identify problem areas which invite follow-up work.

A further feature of the trial referred to above was the multivariate nature of the study. This is common in studies in the rheumatoid arthritis area, where a number of possible related variables are measured. This complicates the issue of handling the withdrawals data if they are to be handled in a multivariate way. For example, Gould points out that his approach may not be feasible when the responses are multivariate. On the other hand such data raise the question of whether multivariate methods can usefully be used when handling clinical trials analyses. The analysis of biochemical laboratory data is another common area in clinical trials where this possibility of using multivariate approaches arises. Despite interest in the pharmaceutical industry in such approaches, little seems to have been done in this area, though some initial work has been done at Kent and again further research would be worthwhile.

6   ASRU consultancy in other areas

Clinical trials work is not the only area of ASRU's industrial liaison activities where tight deadlines are important to the client. This can affect the way a project is planned and carried out. In an ideal situation, which is more likely to exist for projects set up in a conventional way, a logical and thorough progression through the research can be planned. For a research project set up through ASRU, circumstances may dictate that progress by stages is the best compromise possible. Thus, in recent work in collaboration with ICI Plant Protection Division on the population dynamics of the Screech Owl, *Otus asio*, reported by North (1985), in which particular attention was paid to the effect of an additional force of mortality, initial results from the study were required at very short notice. Yet this was an example, like the one described in the early sections, where an extensive preliminary literature search was both appropriate and necessary. Sometimes, of course, the literature search carried out for one project may benefit another later project. Despite the constraints, however, projects of considerable mutual benefit can grow from such inauspicious beginnings. This particular owl modelling work provided input to the student teaching programme at Kent, in the form of a Postgraduate Diploma project (Boddy, 1985) and led to further research (North and Boddy, in preparation). The work was drawn on data collected in an extensive, long-running study of Screech Owls (Van Camp and Henny, 1975) to provide a basis for population modelling of Leslie matrix type, in a deterministic framework, and for a stochastic equivalent.

Not all ASRU work, however, is tied to such tight time constraints. A recent example concerned a study of the effects of environmental stresses on soil microbiological activity, in collaboration with Shell Research Ltd, Sittingbourne. The client in this instance, a biologist who does not claim particular statistical expertise in the area of microbiological field trials, was happy to leave the statistician with an entirely free hand. There was also a realistic amount of time allowed for the work, enabling a fairly thorough analysis of an extensive data set to be carried out. A most rewarding (statistically and biologically) project resulted, in which an array of multivariate methods were applied to a data set of high quality and interest content.

In some of the ASRU consultancy work, it is the objectivity of the analysis, in an issue which is likely to be controversial, which may be sought by the client. The respectability provided by the backing of an academic institution also becomes a factor in such cases. Clearly, in these instances especially the project must be set up in conditions that allow ASRU to be professionally satisfied with its results and conclusions. However, these points are not so different from conditions which might prevail if a client were seeking the advice of an individual academic expert as a consultant.

A prominent feature of ASRU's work over the years, developed through industrial liaison in existence even before the Unit was established, has been the development of user-friendly statistical software. This has resulted in the computer effectively acting as the consultant, as the products have approached the status of expert systems. Work of this type began with the development of interactive computer software in cooperation with ICI plc, Mond Division, to assist scientists in designing their experiments through 'conversation' with the computer: see Jones (1980). Further work to extend this software is desirable, and planned. This was followed by the development of the multiple regression analysis package, U-REG, again developed in conjunction with ICI plc, but now available to anyone, which generated a considerable amount of academic research, culminating in the monograph of Wetherill *et al.* (in preparation).

Work in similar vein, this time supported by the Overseas Development Administration, has led to the development of a user-friendly survey analysis package, U-SP, suitable for running on microcomputers. This work did not arise out of industrial contact, but from a sample surveys course run regularly at Kent for overseas statisticians. However it is anticipated that U-SP should now be able to play a part in industrial applications.

Related research has gone on at Kent in parallel with the development of U-SP. One PhD project involved the study of methods of imputation

(Laurence, in preparation). Whether or not these have a place in an automatic approach to survey analysis is not clear, though the weight of opinion seems to favour the view that in general they do not: more needs to be done on this. A second PhD project, still ongoing, involves the study of model-based approaches to inference from samples surveys, and their comparison with more classical approaches (Wafula, in preparation).

The mention of courses illustrates another role that the Unit is able to play, in conjunction with the statistics department, namely to provide a consultancy service through extra-mural courses. This is becoming an increasingly prominent part of the activities at Kent, but is a commitment which cannot be undertaken lightly, involving as it does intensive expenditure of staff time and effort – often at times when many academics might be thinking of undertaking other activities, such as research. This arises since, by necessity, most extra-mural courses must be run during vacations, though courses for clients on-site can be given more easily during term time.

A final feature of ASRU consultancy work which sets it apart from other forms of university–industry liaison is the question of costing the projects handled. As a self-financing group ASRU clearly needs to make realistic charges for the work it undertakes while also charging at levels considered by its clients to be reasonable. Understandably, most clients prefer to have a scale of charges for a project agreed before its start. This can lead to problems for ASRU. Unit staff can sometimes have difficulties due to unfamiliarity with the in-house routine of a client for handling its data, possibly related to lack of access to the standard software used in-house. More fundamentally, of course, by its very nature, the problems and points of interest of a statistical analysis (which, inevitably, take up more time) only tend to reveal themselves as the analysis progresses, and may not have been anticipated at the stage of setting up a contract for the work. Sometimes more flexible costing arrangements can be employed, which can work to the mutual benefit of both parties. All this is, however, quite far removed from the considerations of collaborative work like that described in the early sections of this chapter, thus completing the spectrum of features relating to statistical liaison with industry from a university group.

Acknowledgement
We are grateful to David Smith for his helpful comments on an earlier version of this paper.

References

Altham, P. M. E. (1978) Two generalisations of the binomial distribution. *Appl.Statist.*, **27**, 162–7.

Aranda-Ordaz, F. J. (1983) An extension of the proportional hazards model for grouped data. *Biometrics*, **39**, 109–17.

Bennett, S. (1983) Log-logistic regression models for survival data. *Appl.Statist.*, **32**, 165–71.

Boddy, A. W. (1985) The stochastic modelling of a Screech Owl population: a computer study. Postgraduate Diploma in Statistics Project Report, University of Kent at Canterbury, 50pp.

Boyce, C. B. C. and Williams, D. A. (1967) The influence of exposure time on the susceptibility of *Australorbis glabratus* to N-tritylmorpholine. *Ann.Trop.Med. Parasitology*, **61** (1), 15–20.

Brooks, R. J. (1984) Approximate likelihood ratio tests in the analysis of beta-binomial data. *Appl.Statist.*, **33**, 285–9.

Carter, E. M. and Hubert, J. J. (1984) A growth-curve model approach to multivariate quantal bioassay. *Biometrics*, **40**, 699–706.

Cox, D. R. (1972) Regression models and life tables (with discussion). *J.R.Statist.Soc.* B, **34**, 187–220.

Diggle, P. J. and Gratton, R. J. (1984) Monte Carlo methods of inference for implicit statistical models (with discussion). *J.R.Statist.Soc.* B, **46**, 193–227.

Dixon, W. J. (1981) *BMDP Statistical Software: User Manual*, University of California Press: Berkeley, 726pp.

Everitt, B. S. (1977) *The Analysis of Contingency Tables*, Chapman and Hall: London.

Farewell, V. T. (1982) The use of mixture models for the analysis of survival data with long-term survivors. *Biometrics*, **38**, 1041–6.

Gladen, B. (1979) The use of the jackknife to estimate proportions from toxicological data in the presence of litter effects. *J.Am.Statist.Ass.*, **74**, 278–83.

Gould, A. L. (1980) A new approach to the analysis of clinical drug trials with withdrawals. *Biometrics*, **36**, 721–7.

Harrison, R. F., Brennan, M., North, P. M., Reed, J. V. and Wickham, E. A. (1984) Is routine episiotomy necessary? *B.Med.J.*, **288**, 1971–5.

Haseman, J. K. and Kupper, L. L. (1979) Analysis of dichotomous response data from certain toxicological experiments. *Biometrics*, **35**, 281–93.

Haseman, J. K. and Soares, E. R. (1976) The distribution of fetal death in control mice and its implications on statistical tests for dominant-lethal effects. *Mutat.Res.* **41**. 277–88.

James, D. A. and Smith, D. M. (1982) Analysis of results from a collaborative study of the dominant lethal assay. *Mutat.Res.*, **97**, 303–14.

Jarrett, R. G. (1984) A look at the MICE (mortality index for chicken embryos) test, CSIRO Division of Mathematics and Statistics, Report VT 84/16.

Jones, B. (1980) The computer as a statistical consultant. *Bull.Appl.Statist.*, **7**, 168–95.

Jones, B., Morgan, B. J. T. and Wetherill, G. B. (1983) Building bridges between the academic and real worlds. In D. R. Grey *et al.* (eds.) *Proceedings of the First International Conference on Teaching Statistics*, Teaching Statistics Trust, pp. 521–6.

Kalbfleish, J. D., Krewksi, D. R. and Van Ryzin, J. (1983) Dose-response models for time-to-response toxicity data. *Can.J.Statist.*, **11**, 25–49.

Kleinman, J. C. (1973) Proportions with extraneous variance: single and independent samples. *J.Am.Statist.Ass.*, **68**, 46–54.

Kooijman, S. A. L. M. (1981) Parametric analyses of mortality rates in bioassays. *Wat.Res.*, **15**, 105–19.

Kupper, L. L. and Haseman, J. K. (1978) The use of a correlated binomial model for the analysis of certain toxicological experiments. *Biometrics*, **35**, 281–93.

McCullagh, P. (1980) Regression models for ordinal data. *J.R.Statist.Soc.* B, **42**, 109–42.

North, P. M. (1985) A computer modelling study of the population dynamics of the Screech Owl (*Otus asio*). *Ecol. Modelling*, **30**, 105–43.

Pack, S. E. (1985) A note on the equivalence of two estimators for over-dispersed proportions. Unpublished manuscript.

Pack, S. E. (1986*a*) Hypothesis testing for proportions with over-dispersion. To appear in *Biometrics*.

Pack, S. E. (1986*b*) A comparison of binomial generalizations for toxicological data. To appear in *Biometrics*.

Paul, S. R. (1982) Analysis of proportions of affected foetuses in teratological experiments. *Biometrics*, **38**, 361–70.

Paul, S. R. (1984) A three-parameter generalization of the binomial distribution. Submitted for publication.

Peto, R., Pike, M. C., Armitage, P., Breslow, N. E., Cox, D. R., Howard, S. V., Mantel, N., McPherson, K., Peto, J. and Smith, P. G. (1977) Design and analysis of randomized clinical trials requiring prolonged observation of each patient. II. Analysis and examples. *Br.J.Cancer*, **35**, 1–39.

Pregibon, D. (1981) Logistic regression diagnostics. *Annls Statist.* **9**, 705–24.

Pregibon, D. (1982) Resistant fits for some commonly used logistic models with medical applications. *Biometrics*, **38**, 485–98.

Prentice, R. L. (1986) Binary regression using an extended Beta-binomial distribution, with discussion of correlation induced by covariate measurement errors. To appear in *J.Am.Statist.Ass.*

Puri, P. S. and Senturia, J. (1972) On a mathematical theory of quantal response assays. *Proceedings of the 6th Berkeley Symposium on Mathematical Statistics*, University of California Press: Los Angeles.

Shirley, E. A. C. and Hickling, H. (1981) An evaluation of some statistical methods for analysing numbers of abnormalities found amongst litters in teratology studies. *Biometrics*, **37**, 819–29.

Silvey, S. D. (1975) *Statistical Inference*, Chapman & Hall: London.

Smith, D. M. (1983) AS 189 – Maximum likelihood estimation of the parameters of the beta binomial distribution. *Appl.Statist.*, **32**, 196–204.

Van Camp, L. T. and Henny, C. J. (1975) The Screech Owl, its life history and population ecology in Northern Ohio. United States Department of the Interior Fish and Wildlife Service. North American Fauna. Report No. 71.

Vuataz, L. and Sotek, J. (1978) Use of the beta-binomial distribution for 'weak mutagenic activity' (Part 2). *Mutat.Res.*, **52**, 211–30.

Williams, D. A. (1975) The analysis of binary responses from toxicological experiments involving reproduction and teratogenicity. *Biometrics*, **31**, 949–52.

Williams, D. A. (1982) Extra binomial variation in logistic linear models. *Appl.Statist.*, **31**, 144–8.

Wolynetz, M. S. (1979) AS 139 – Maximum likelihood estimation in a linear model from confined and censored normal data. *Appl.Statist.*, **28**, 195–206.

11

Inspection for faulty components before or after assembly of manufactured items

P. M. E. ALTHAM

Summary

A problem in quality control, supplied by Marks and Spencer plc in the particular context of garment manufacture, is described. A fixed number of components is to be assembled to produce a finished item in the factory. Each component has a small probability of being faulty, and any component can be inspected before the item is assembled, if necessary being replaced by a perfect component. The finished item is checked for faults at the factory, and depending on the outcome of this checking may be sent to the store as perfect, or sold as a second, or repaired. A BASIC program is described, whose purpose is to help the manufacturer to decide which components, if any, should be inspected before assembly, and to illustrate the process by a simulation. The optimisation and statistical inference aspects of the problem are discussed.

Preamble

I am sure most university statisticians find that the experience of statistical consulting is essential to their professional well-being and sanity, as well as often being great fun, affording a glimpse into other worlds besides their own department or university. Like many statisticians, I find 'research' and 'consulting' are not readily disentangled, and indeed to separate one's activities into these two areas is probably rather unhealthy. However, there is obviously a danger that if one's research is too much problem-driven, one may be subject to a series of random impetuses, so that the net direction may be rather unclear. To guard against this, we all need to be able to stand back from time to time and reflect on our rather random collection of problems; to work out which, if any, are worth pursuing further, to work out whether there are any connections, possibly worth exploiting, between these problems. The advantage of keeping an open mind about the type of problem one is prepared to handle is that one is then very positively stimulated to try to keep up with one's subject in a broad sense, rather than just plough one or two very deep furrows.

I found the experience of consulting described in this chapter very educational, even if (not very surprisingly) there were no new theorems, earth-shattering or otherwise, to be proved. My colleagues at Marks and Spencer rightly insisted that a proper study of the problem meant that I must visit several factories, to see various types of garment inspection in operation. Having the manufacturing process explained to me over the noise of the machines, and over the noise of Radio 1, was a fairly rude, though not at all unwelcome, contrast to the quiet of my university office! While I found the visits to the factories and to Marks and Spencer's Head Office immensely interesting, sitting in the quiet of my office with pencil, paper and computer was of course a very substantial part of the enterprise. My first attempts to pose and solve the problem, in the language of mathematics/probability, did not go down at all well with my clients and this, I must confess, left me temporarily a little dashed; I had been quite proud of my elegant, Cambridge-style formulation. I have spent many years emphasising to my students the importance of good communication with one's clients, and at one stage in this project it looked as if I myself was in danger of failing rather badly in this respect. However, at this point the microcomputer (in spite of its reputation as having a great potential for time-wasting) came to the rescue, and it's still hard for me to think of any better way to present the results, particularly to people who are not all that interested in the mathematics but just want a simple numerical answer. I am obviously not a professional programmer, but simply use the programs as a means to an end. Examples of the output of my programs are given at the end of the chapter; since the main body of the chapter was written, the format of the output has been modified somewhat, following comments from potential users. Probably a programming expert could have some fun revamping the programs into a slicker form. For me the programming exercise played a very important role in familiarising myself with my then new microcomputer and BASIC. Mainframe computing would have been too unwieldy for this problem; I needed to be able to demonstrate the results on a portable computer, and on a computer that is easily available and affordable by factories in the UK.

Since the main body of the chapter was written, Marks and Spencer have generously agreed that the programs may be made more widely available, for the financial benefit of Cambridge University, so steps are being taken with the help of the University's Wolfson Industrial Unit to promote the work.

1   Introduction

This problem was supplied by Marks and Spencer plc, and I am very grateful to them for the opportunity to work on an interesting and worthwhile question. For me the problem arose in the particular context of quality control of fabric faults in garments made from woven fabrics. However, the general problem of how best to inspect for faults in a manufactured item consisting of a number of individual components, each of which is prone to faults, and when no inspection is entirely error-free, is of course very widespread. As a problem in quality control it may have been tackled before, but if so I am unaware of it.

To revert to a particular example of this problem, consider the manufacture of a shirt. The fabric supplied to the factory is cut into the shirt components (termed 'panels') and, because the fabric is never quite perfect, there is a slight chance that one or more of the panels for a given shirt may have a fabric fault. The manufacturer has a number of options open to him. He can arrange for all, some or none of the panels to be inspected for fabric faults before the shirt is sewn, and then any panels found to be faulty could be replaced by perfect ones. The panels are then sewn together to form the shirt, and the finished item is given a final inspection. At this stage there are several possibilities:

   (i) the garment is perfect and is sent to the Marks and Spencer store as such;

   (ii) it is found by the factory to be faulty, and is sold as a 'second';

   (iii) it is found to be faulty, repaired at the factory, and then sold;

   (iv) it is faulty, but sent in error to the store.

In the examples on which I worked, this last event had only a very small chance of occurring but, because of the huge total numbers of garments produced, even a rare event is not economically negligible.

(In the formal statement and solution of the problem that follows, possibility (iii) is not taken into account, but given the relevant extra information the solution could easily be generalised to remedy this.)

My problem then was the following: what advice can a statistician give to the manufacturers about the best methods of inspection for fabric faults? Since the problem is not confined to the manufacture of shirts, nor even to the manufacture of garments, it will be discussed in a rather general framework.

2   Formal statement of the problem

The *item* manufactured consists of n *components*, which we call $a_1, ..., a_n$. Some of these may be faulty when supplied to the factory. Initially we consider the following two methods of inspection for faults in the finished item. (For simplicity, faults incurred *during* assembly of components are ignored: they would affect both schemes equally.)

Scheme 1: component inspection

Before the assembly of an item, each component $a_1, ..., a_n$ is checked, and if found to be faulty, replaced by a perfect component. The cost of replacing a faulty a_i is £p_i. *After* assembly, the whole item is checked: if any component is found to be faulty the item is sold as a 'second'; otherwise the item is sold to the customer at the full price. If the customer subsequently finds he has been sold a faulty item, then the factory incurs a financial penalty.

Scheme 2: no component inspection

This is as in scheme 1, but without the inspection of individual components before assembly. Thus this scheme is cheaper in terms of inspection costs, but could result in more faulty items being produced.

3   Notation and solution

Let $\pi_i = \text{Pr} \, (a_i$ has one or more faults$)$ $i = 1, ..., n$. (Usually π_i would be small, $\pi_i < 0.05$ say.)

In the case where a_i is a panel of fabric of area A_i, and fabric faults occur in a Poisson process, rate λ per unit area of fabric, then we would find that $\pi_i = 1 - \exp(-\lambda A_i)$, so that for λA_i small, $\pi_i \approx \lambda A_i$.

Assume that if the factory sends a perfect item to the customer, it receives an amount £P; if it sells a faulty item as a 'second' it receives an amount £S, and if it has the misfortune to sell a faulty item as a perfect one, it incurs a penalty £F. Clearly in most practical cases we would find $P > S$, though in some cases the difference between P and S is not substantial.

Note that the net penalty F may be negative for the following reason. If the factory supplies a faulty item to the customer, and the customer returns it to the factory, then the factory generally has to pay the transport cost, but may subsequently be able to sell the item as a second, in which case F may be negative. However, if the faulty item has to be scrapped, then F will certainly be positive.

In what follows we make no assumptions about the manufacturer's *profit*; we work only in terms of the (expected) amount of money received by the manufacturer per item. Manufacturers may understandably not wish to divulge their desired profit per item, and this quantity is not required in the quantitative assessment of the scheme for detecting faults in the components.

The problems arise here because, however well-trained and careful the inspectors are, they can never be completely perfect; they are bound to miss faulty components or faulty finished items occasionally. Thus, let

$$\alpha_F = \text{Pr (faulty item noted as faulty at final inspection)}$$
$$\beta_F = \text{Pr (faulty item missed at final inspection)},$$

so that $\alpha_F + \beta_F = 1$ (and generally α_F would be large, say > 0.95). We ignore the possibility that a perfect item is wrongly noted as faulty; similarly for perfect components.

Finally let

$$\alpha_i = \text{Pr (a faulty component } a_i \text{ is noted as faulty at initial}$$
$$\text{inspection for scheme 1)}$$

and let

$$\beta_i = \text{Pr (a faulty component } a_i \text{ is missed in scheme 1)}$$
$$= 1 - \alpha_i, i = 1, ..., n,$$

and let £C be cost of inspection of the n components before assembly in scheme 1. Presumably C is a function of $\alpha_1, ..., \alpha_n$.

We are now in a position to compare schemes 1 and 2 by deriving the amount received by the manufacturer per item produced, say £Y_i for scheme i. (In practice establishing numerical values for the above quantities $\pi_i, ..., \beta_i, C$ may not be straightforward.) From a very simple-minded point of view, we would then say that scheme 2 is cheaper than scheme 1 if $E(Y_2) > E(Y_1)$: that is, if the expected amount *received* per item is larger for scheme 2 than for scheme 1. However, the manufacturer may well have other criteria in mind besides expected cost, and so we may like to take a closer look at the difference between the two schemes by comparing the *distributions* of Y_1 and Y_2, if necessary by simulation. Furthermore, from the point of view of the customer's goodwill, the manufacturer will want to know the probability that an item sold to the customer as perfect actually *is* perfect. This probability for scheme 1 is clearly going to be greater than the corresponding probability for scheme 2.

We start with *scheme 2*, since it is the simpler one. Consider the manufacture of a particular item. Define the indicator random variables,

$$A_i = 0 \quad \text{if the } i\text{th component is perfect}$$
$$A_i = 1 \quad \text{if the } i\text{th component is faulty}$$
$$\left.\begin{array}{c}\end{array}\right\} \quad \text{thus} \quad E(A_i) = \pi_i, \quad i = 1, ..., n.$$

Put $Z_2 = (1-A_1), \ldots, (1-A_n)$, thus

$Z_2 = 1$ if and only if all components are perfect,

$Z_2 = 0$ otherwise.

If $Z_2 = 0$, then a faulty item will be produced, so what we then need to worry about is whether the final inspection after assembly picks it up. Hence, if $Z_2 = 0$, define $I_S = 1$ if the secondary inspection finds the item faulty, $I_S = 0$ otherwise. (Note there is a certain oversimplification here, since presumably the *more* faulty components in the finished item, the *higher* the probability that the inspector finds the item faulty; we are assuming this probability is constant.) Thus we simply take $E(I_S \mid Z_2 = 0)$ as α_F. We see that

$$Y_2 = +P \quad \text{if} \quad Z_2 = 1,$$
$$Y_2 = +S \quad \text{if} \quad Z_2 = 0 \quad \text{and} \quad I_S = 1,$$
$$Y_2 = -F \quad \text{if} \quad Z_2 = 0 \quad \text{and} \quad I_S = 0,$$

from which it follows that

$$Y_2 = PZ_2 + (1-Z_2)\,(I_S\,S - F(1-I_S))$$

Hence

$$E(Y_2) = P\delta_2 + (1-\delta_2)\,(\alpha_F S - F\beta_F)$$

where $\delta_2 = E(Z_2) = (1-\pi_1) \ldots (1-\pi_n)$.

Now consider the manufacture of a particular item under scheme 1. Take the indicator variables A_i as before, and define new variables B_i as follows:

if $A_i = 0$ define $B_i = 1$,

if $A_i = 1$ define $B_i = 1$ with probability α_i,

$B_i = $ with probability $1 - \alpha_i$.

Thus, for component a_i, there are three possibilities:

(i) it is initially perfect, in which case $A_i = 0$ and $B_i = 1$,

(ii) it is initially faulty, and the preliminary inspection finds the fault, in which case $A_i = 1$, $B_i = 1$, and a cost ρ_i of replacement is incurred,

(iii) it is initially faulty but preliminary inspection misses the fault, in which case $A_i = 1$, $B_i = 0$, and there is no replacement cost.

Thus a perfect item is assembled if and only if $B_1 \ldots B_n = 1$; define $Z_1 = B_1 \ldots B_n$. Define I_S as the outcome of the secondary inspection as before; thus, given $Z_1 = 0$, take

$$I_S = 1 \text{ with probability } \alpha_F,$$
$$I_S = 0 \text{ with probability } \beta_F.$$

Hence the reader may verify that

$$Y_1 = P - \sum_1^n p_i A_i B_i - C, \quad \text{if} \quad Z_1 = 1,$$

$$Y_1 = S - \sum_1^n p_i A_i B_i - C, \quad \text{if} \quad Z_1 = 0 \quad \text{and} \quad I_S = 1,$$

$$Y_1 = -F - \sum_1^n p_i A_i B_i - C, \quad \text{if} \quad Z_1 = 0 \quad \text{and} \quad I_S = 0.$$

Hence we may write

$$Y_1 = -C - \sum p_i A_i B_i + PZ_1 + (1 - Z_1)(+SI_S - F(1 - I_S))$$

where $(A_1, B_1), \ldots, (A_n, B_n)$ are independent, and

$$E(A_i) = \pi_i, \; E(B_i | A_i = 1) = \alpha_i, \; E(B_i | A_i = 0) = 1,$$

and

$$E(I_S | Z_1 = 0) = \alpha_F, \; E(Z_1) = \delta_1 \text{ say.}$$

Hence

$$E(A_i B_i) = \pi_i \alpha_i,$$

and

$$E(Y_1) = -C - \sum p_i \alpha_i \pi_i + P\delta_1 + (1 - \delta_1)(+S\alpha_F - F\beta_F).$$

The quantity δ_i is the probability that scheme i produces a perfect item:

$$| \quad \delta_1 = \prod_1^n (1 - \pi_i \beta_i)$$

and

$$\delta_2 = \prod_1^n (1 - \pi_i)$$

so that clearly $\delta_1 > \delta_2$, as we would expect.

For example, with $n = 10$, $\pi_i = 0.05$ for all i and $\beta_i = 0.01$ for all i, we find $\delta_1 = 0.995, \delta_2 = 0.599$.

The probability that an item supplied to the customer as perfect is actually perfect, with scheme 1, is $\delta_i / [\delta_i + (1 - \delta_i) \beta_F]$, which will clearly be a little higher than δ_i.

4   Discussion of results, and presentation

Schemes 1 and 2 are two *extremes* in possible methods of inspection. Note that with α_i as the accuracy of inspection of the ith panel in scheme 1, $\alpha_i > 0$ for $i = 1, \ldots, n$ corresponds to the inspection of *every* panel before assembly, but the manufacturer may actually prefer to take $\alpha_i > 0$ for some values of i, and $\alpha_i = 0$ for the remaining i, depending on the fault rates π_i and replacement costs p_i for the individual panels. Of course if he takes $\alpha_i = 0$ for all i, then scheme 1 reduces to scheme 2, the case of no panel inspection.

What the manufacturer really wants to know from the statistician, in deciding on the best method of inspection, is the numerical value of

$$E(Y_1) = P\delta_1 + (1-\delta_1)(S\alpha_F - F\beta_F) - C(\alpha_1, \dots, \alpha_n)$$
$$-\sum_1^n \rho_i \alpha_i \pi_i, \qquad (4.1)$$

where $\delta_1 = \prod_1^n (1 - \pi_i + \pi_i \alpha_i)$.

The most effective way of presenting the results of (4.1) was to use a microcomputer program. The formula involves too many parameters for a simple graphical presentation or book of tables to be a practical possibility, so I wrote a program in BBC BASIC. This had the great advantage that, using the nice notation of the indicator variables described in Section 3, I was able to demonstrate the validity of (4.1) by simulation. This is especially suitable for the layman, who probably is unfamiliar with terms such as 'expectation', but finds long-run convergence to the expected value quite convincing.

Examples of the output from the two relevant programs are given in Tables 1 and 2. Note that the examples are for very small scale problems, with rather unrealistically high fault rates, and only rather small numbers of simulations, for reasons of economy of space in this chapter. The programs are designed to be as user-friendly as possible. The user simply types in the *values* of the appropriate parameters after each question. The program deals with a slightly more general problem than those given in schemes 1 and 2 above, to allow for the fact that not all faulty items sent to the store will be detected and returned to the factory. Thus the user is asked to specify the rate of detection of the faulty item at the store; this would tend to be higher for an expensive item than a cheaper item. The outcome of the simulations in Table 1 shows the state of the panels in brackets, 'OK' or 'DUD', and following the brackets, the state of the whole item, again 'OK' or 'DUD'. In Table 2, which shows the outcome of the program demonstrating panel inspection, the *initial* state of the panel is shown in the brackets as 'OK' or 'DUD', and its state *after* panel inspection is shown as 'ok' or 'dud'. The state of the whole item, following panel inspection, is shown after the brackets as 'OK' or 'DUD'. The last figure for each simulation shows the amount received by the manufacturer for that particular item, allowing for panel replacement if that occurred.

For listings of the programs please write to the author.

Table 1

INSPECTION FOR FABRIC FAULTS, WITH NO PANEL INSPECTION PRIOR
TO ASSEMBLY OF GARMENT
Copyright P. M. E. Altham 1985 for Marks and Spencer plc

If you make an error in what follows, you can probably correct it by the DELETE key.
Otherwise just press ESCAPE & then type RUN
Do not press BREAK

IF YOU WANT EACH SIMULATION PRINTED OUT, ANSWER 1, ELSE 0?   1
IF FACTORY SENDS PERFECT ITEM TO STORE, IT GETS?   10
IF FACTORY SELLS FAULTY ITEM AS A SECOND, IT GETS?   6
IF FACTORY SENDS FAULTY ITEM TO STORE, IT INCURS PENALTY? -3
NUMBER OF PANELS?   4

Now give fault rate for each panel, between 0 and 1

    1   FAULT RATE FOR THIS PANEL?   0.01
    2   FAULT RATE FOR THIS PANEL?   0.01
    3   FAULT RATE FOR THIS PANEL?   0.02
    4   FAULT RATE FOR THIS PANEL?   0.1

Now give accuracy of final inspection at factory, between 0 and 1
ACCURACY OF FINAL INSPECTION?   0.8

Now give rate of detection of faulty item at store, between 0 and 1
DETECTION?   0.5

PERFECT ITEMS SENT TO STORE   86.445%
SECONDS PRODUCED, & DETECTED AT FACTORY   10.844%
FAULTY ITEMS SENT TO STORE   2.711%
RTMS† FROM STORE   1.356 as % of all items made

This means that of the items sent to the store, we expect a proportion to be returned to
factory of   1.520%

EXPECTED AMOUNT RECEIVED PER ITEM   9.471

Percentage loss on contract due to rtms and seconds   5.287%

NUMBER OF SIMULATIONS?   10

(DUD	DUD	DUD	DUD)	DUD, FAULT SPOTTED, SECOND	6.000
(OK	OK	DUD	DUD)	DUD, FAULT SPOTTED, SECOND	6.000
(OK	OK	OK	OK)	OK, PERFECT ITEM	10.000
(OK	OK	OK	DUD)	DUD, FAULT SPOTTED, SECOND	6.000
(OK	OK	OK	OK)	OK, PERFECT ITEM	10.000
(OK	OK	OK	OK)	OK, PERFECT ITEM	10.000
(OK	OK	OK	OK)	OK, PERFECT ITEM	10.000
(OK	OK	OK	OK)	OK, PERFECT ITEM	10.000
(OK	OK	OK	DUD)	DUD, FAULT SPOTTED, SECOND	6.000
(OK	OK	OK	OK)	OK, PERFECT ITEM	10.000

OUTCOME OF SIMULATIONS

MEAN RECEIVED PER ITEM   8.400, SE   2.066
Here we assume that any faulty item sent to the store will be an rtm
NUMBER OF SIMULATIONS   10.000
NUMBER OF PERFECTS   6.000   60.000%
NUMBER OF SECONDS   4.000   40.000%
FAULTY GOODS TO STORE   0.000   0.000%

---

† RTM and rtm are abbreviations for 'returned to manufacturer'.

Table 1 (*cont.*)

If you want more simulations with same parameters type GOTO 290(return)
If you want to change the parameters type RUN(return)
>RUN
INSPECTION FOR FABRIC FAULTS, WITH NO PANEL INSPECTION PRIOR
TO ASSEMBLY OF GARMENT
Copyright P. M. E. Altham 1985 for Marks and Spencer plc

If you make an error in what follows, you can probably correct it by the DELETE key.
Otherwise just press ESCAPE & then type RUN
Do not press BREAK

IF YOU WANT EACH SIMULATION PRINTED OUT, ANSWER 1, ELSE 0?　　0
IF FACTORY SENDS PERFECT ITEM TO STORE, IT GETS?　　12
IF FACTORY SELLS FAULTY ITEM AS A SECOND, IT GETS?　　0
IF FACTORY SENDS FAULTY ITEM TO STORE, IT INCURS PENALTY?　　2
NUMBER OF PANELS?　　6

Now give fault rate for each panel, between 0 and 1

1	FAULT RATE FOR THIS PANEL?	0.1
2	FAULT RATE FOR THIS PANEL?	0.1
3	FAULT RATE FOR THIS PANEL?	0.01
4	FAULT RATE FOR THIS PANEL?	0.02
5	FAULT RATE FOR THIS PANEL?	0.05
6	FAULT RATE FOR THIS PANEL?	0.001

Now give accuracy of final inspection at factory, between 0 and 1
ACCURACY OF FINAL INSPECTION?　　0.8
Now give rate of detection of faulty item at store, between 0 and 1
DETECTION?　　1

PERFECT ITEMS SENT TO STORE　　74.582%
SECONDS PRODUCED, & DETECTED AT FACTORY　　20.334%
FAULTY ITEMS SENT TO STORE　　5.084%
RTMS FROM STORE　　5.084 as % of all items made

This means that of the items sent to the store, we expect a proportion to be returned to factory of　　6.381%

EXPECTED AMOUNT RECEIVED PER ITEM　　8.848

Percentage loss on contract due to rtms and seconds　　26.265%

NUMBER OF SIMULATIONS?　　10

OUTCOME OF SIMULATIONS

MEAN RECEIVED PER ITEM　　12.000, SE　　0.000
Here we assume that any faulty item sent to the store will be an rtm
NUMBER OF SIMULATIONS　　10.000

NUMBER OF PERFECTS　　10.000　　100.000%
NUMBER OF SECONDS　　0.000　　0.000%
FAULTY GOODS TO STORE　　0.000　　0.000%

If you want more simulations with same parameters type GOTO 290(return)
If you want to change the parameters type RUN(return)

Table 2

INSPECTION FOR FABRIC FAULTS, WITH PANEL INSPECTION BEFORE
FINAL ASSEMBLY OF GARMENT

Copyright P. M. E. Altham 1985 for Marks and Spencer plc
If you make an error in what follows, you can probably correct it by DELETE key.
Otherwise just press ESCAPE & then type RUN
Do not press BREAK

IF YOU WANT EACH SIMULATION PRINTED OUT, ANSWER 1, ELSE 0?   1
IF FACTORY SENDS PERFECT ITEM TO STORE, IT GETS?   10
IF FACTORY SENDS FAULTY ITEM AS A SECOND, IT GETS?   8
IF FACTORY SENDS FAULTY ITEM TO STORE, IT INCURS PENALTY?   2
NUMBER OF PANELS?   6

Give fault rate between 0 and 1

1	FAULT RATE FOR THIS PANEL?	0.01
2	FAULT RATE FOR THIS PANEL?	0.1
3	FAULT RATE FOR THIS PANEL?	0.01
4	FAULT RATE FOR THIS PANEL?	0.01
5	FAULT RATE FOR THIS PANEL?	0.05
6	FAULT RATE FOR THIS PANEL?	0.02

1	REPLACEMENT COST OF THIS PANEL?	0.5
2	REPLACEMENT COST OF THIS PANEL?	0.5
3	REPLACEMENT COST OF THIS PANEL?	0.5
4	REPLACEMENT COST OF THIS PANEL?	0.6
5	REPLACEMENT COST OF THIS PANEL?	1
6	REPLACEMENT COST OF THIS PANEL?	0.7

Give accuracy rate between 0 and 1 (1 corresponds to perfect inspection, 0 to no
inspection)
NB If you set accuracy of inspection of a panel to 0, then in effect this panel is not
inspected prior to assembly

1	ACCURACY OF INSPECTION OF THIS PANEL?	0
2	ACCURACY OF INSPECTION OF THIS PANEL?	0.99
3	ACCURACY OF INSPECTION OF THIS PANEL?	0.8
4	ACCURACY OF INSPECTION OF THIS PANEL?	0.8
5	ACCURACY OF INSPECTION OF THIS PANEL?	0
6	ACCURACY OF INSPECTION OF THIS PANEL?	0.90

ACCURACY OF FINAL INSPECTION?   0.95

Now give rate of detection of faulty item at store, between 0 and 1

DETECTION?   1

PERFECT ITEMS SENT TO STORE   93.393%, PRICE 10.000 EACH
SECONDS PRODUCED, & DETECTED AT FACTORY   6.276%,
PRICE 8.000 EACH
FAULTY ITEMS SENT TO STORE   0.330%

RTMS FROM STORE   0.330 as % of all items made; COST 2.000 EACH

This means that, of the items sent to the store, we expect a proportion to be returned to
the factory of 0.352%

EXPECTED AMOUNT RECEIVED PER ITEM   9.764

Table 2 (*cont.*)

Percentage loss on contract due to seconds & rtms   2.418%
(NB This does not take account of the cost of the panel inspection itself; this cost would presumably depend on the accuracy of the panel inspection)

SIMULATIONS, where for simplicity we assume any faulty item sent to the store will be an rtm

NUMBER OF SIMULATIONS?   15
(OK   ok   DUD   ok   OK   ok   OK   ok   OK   ok   OK   ok)   OK,
      PERFECT ITEM   9.500

(OK   ok   OK   ok   OK   ok   OK   ok   OK   ok   OK   ok)   OK,
      PERFECT ITEM   10.000

(OK   ok   OK   ok   OK   ok   OK   ok   OK   ok   OK   ok)   OK,
      PERFECT ITEM   10.000

(DUD   dudOK   ok   OK   ok   OK   ok   OK   ok   OK   ok)   DUD,
FAULT SPOTTED, SECOND   8.000

(OK   ok   OK   ok   OK   ok   OK   ok   OK   ok   OK   ok)   OK,
      PERFECT ITEM   10.000

(OK   ok   OK   ok   OK   ok   OK   ok   OK   ok   OK   ok)   OK,
      PERFECT ITEM   10.000

(OK   ok   OK   ok   OK   ok   OK   ok   OK   ok   OK   ok)   OK,
      PERFECT ITEM   10.000

(OK   ok   OK   ok   OK   ok   OK   ok   DUD   dud OK   ok)   DUD,
FAULT SPOTTED, SECOND   8.000

(OK   ok   OK   ok   OK   ok   OK   ok   OK   ok   OK   ok)   OK,
      PERFECT ITEM   10.000

(OK   ok   OK   ok   OK   ok   OK   ok   OK   ok   OK   ok)   OK,
      PERFECT ITEM   10.000

(OK   ok   OK   ok   OK   ok   OK   ok   OK   ok   OK   ok)   OK,
      PERFECT ITEM   10.000

(OK   ok   OK   ok   OK   ok   OK   ok   OK   ok   OK   ok)   OK,
      PERFECT ITEM   10.000

(OK   ok   OK   ok   OK   ok   OK   ok   OK   ok   OK   ok)   OK,
      PERFECT ITEM   10.000

(OK   ok   OK   ok   OK   ok   OK   ok   OK   ok   OK   ok)   OK,
      PERFECT ITEM   10.000

(OK   ok   OK   ok   OK   ok   OK   ok   OK   ok   OK   ok)   OK,
      PERFECT ITEM   10.000

OUTCOME OF SIMULATIONS

MEAN RECEIVED PER ITEM   9.700, SE   0.702
NUMBER OF SIMULATIONS   15.000
NUMBER OF PERFECTS   13.000   86.667%
NUMBER OF SECONDS   2.000   13.333%
NUMBER OF RTMS   0.000   0.000%

Table 2 (*cont.*)

If you want more simulations with the same parameters, type GOTO 370 (return)
If you want to vary accuracy of panel inspection, type GOTO 270
>GOTO 270
    1.000   ACCURACY OF INSPECTION OF THIS PANEL?   0.99
    2.000   ACCURACY OF INSPECTION OF THIS PANEL?   0.99
    3.000   ACCURACY OF INSPECTION OF THIS PANEL?   0.99
    4.000   ACCURACY OF INSPECTION OF THIS PANEL?   0.99
    5.000   ACCURACY OF INSPECTION OF THIS PANEL?   0.99
    6.000   ACCURACY OF INSPECTION OF THIS PANEL?   0.99
ACCURACY OF FINAL INSPECTION?   0.95
Now give rate of detection of faulty item at store, between 0 and 1
DETECTION?   1

    PERFECT ITEMS SENT TO STORE   99.800%, PRICE 10.000 EACH
    SECONDS PRODUCED, & DETECTED AT FACTORY   0.190%,
    PRICE 8.000 EACH
    FAULTY ITEMS SENT TO STORE   0.010%

RTMS FROM STORE   0.010 as % of all items made; COST 2.000 EACH

This means that, of the items sent to the store, we expect a proportion to be returned to the factory of 0.010%

EXPECTED AMOUNT RECEIVED PER ITEM   9.866
Percentage loss on contract due to seconds & rtms   1.355%
(NB   This does not take account of the cost of the panel inspection itself; this cost would presumably depend on the accuracy of the panel inspection)

SIMULATIONS, where for simplicity we assume any faulty item sent to the store will be an rtm
NUMBER OF SIMULATIONS?   15

(OK ok OK ok OK ok   OK ok   OK ok   OK ok)   OK,
        PERFECT ITEM   10.000

(OK ok OK ok OK ok   OK ok   OK ok   OK ok)   OK,
        PERFECT ITEM   10.000

(OK ok OK ok OK ok   OK ok   OK ok   OK ok)   OK,
        PERFECT ITEM   10.000

(OK ok OK ok OK ok   OK ok   OK ok   OK ok)   OK,
        PERFECT ITEM   10.000

(OK ok OK ok OK ok   OK ok   OK ok   OK ok)   OK,
        PERFECT ITEM   10.000

(OK ok OK ok OK ok   OK ok   DUD ok OK ok)   OK,
        PERFECT ITEM   9.000

(OK ok DUD ok OK ok   OK ok   OK ok   OK ok)   OK,
        PERFECT ITEM   9.500

(OK ok OK ok OK ok   OK ok   DUD ok OK ok)   OK,
        PERFECT ITEM   9.000

(OK ok OK ok OK ok   OK ok   OK ok   OK ok)   OK,
        PERFECT ITEM   10.000

Table 2 (cont.)

(OK   ok   OK   ok   OK   ok   OK   ok   OK   ok   OK   ok)   OK,
   PERFECT ITEM   10.000

(OK   ok   OK   ok   OK   ok   OK   ok   OK   ok   OK   ok)   OK,
   PERFECT ITEM   10.000

(OK   ok   DUD   ok   OK   ok   OK   ok   OK   ok   OK   ok)   OK,
   PERFECT ITEM   9.500

(OK   ok   OK   ok   OK   ok   OK   ok   OK   ok   OK   ok)   OK,
   PERFECT ITEM   10.000

(OK   ok   OK   ok   OK   ok   OK   ok   OK   ok   OK   ok)   OK,
   PERFECT ITEM   10.000

(OK   ok   OK   ok   OK   ok   OK   ok   OK   ok   OK   ok)   OK,
   PERFECT ITEM   10.000

OUTCOME OF SIMULATIONS

MEAN RECEIVED PER ITEM 9.800, SE 0.368
NUMBER OF SIMULATIONS   15.000
NUMBER OF PERFECTS   15.000 100.000%
NUMBER OF SECONDS   0.000 0.000%
NUMBER OF RTMS   0.000 0.000%
If you want more simulations with the same parameters, type GOTO 370 (return)
If you want to vary accuracy of panel inspection, type GOTO 270

>GOTO 270
    1.000   ACCURACY OF INSPECTION OF THIS PANEL? 0
    2.000   ACCURACY OF INSPECTION OF THIS PANEL? 0
    3.000   ACCURACY OF INSPECTION OF THIS PANEL? 0
    4.000   ACCURACY OF INSPECTION OF THIS PANEL? 0
    5.000   ACCURACY OF INSPECTION OF THIS PANEL? 0
    6.000   ACCURACY OF INSPECTION OF THIS PANEL? 0
ACCURACY OF FINAL INSPECTION? 0.95
Now give rate of detection of faulty item at store, between 0 and 1
DETECTION? 1

 PERFECT ITEMS SENT TO STORE 81.301%, PRICE 10.000 EACH
 SECONDS PRODUCED, & DETECTED AT FACTORY 17.764%,
 PRICE 8.000 EACH
 FAULTY ITEMS SENT TO STORE 0.935%

RTMS FROM STORE 0.935 as % of all items made; COST 2.000 EACH

This means that of the items sent to the store, we expect a proportion to be
 returned to the factory of 1.137%

EXPECTED AMOUNT RECEIVED PER ITEM 9.533
Percentage loss on contract due to seconds & rtms 4.904%
(NB   This does not take account of the cost of the panel inspection itself; this cost would
presumably depend on the accuracy of the panel inspection)

Table 2 (*cont.*)

SIMULATIONS, where for simplicity we assume any faulty item sent to the store will be an rtm

NUMBER OF SIMULATIONS?   15

| (OK | ok | OK | ok | OK | ok | OK | ok | DUD | dud | OK | ok) | DUD, |

FAULT SPOTTED, SECOND   8.000

(OK ok OK ok OK ok   OK ok   OK ok   OK ok) OK,
       PERFECT ITEM   10.000

(OK ok OK ok OK ok   OK ok   OK ok   OK ok) OK,
       PERFECT ITEM   10.000

(OK ok OK ok OK ok   OK ok   OK ok   OK ok) OK,
       PERFECT ITEM   10.000

(OK ok DUD dud OK ok   OK ok   OK ok   OK ok) DUD,
       FAULT SPOTTED, SECOND   8.000

(OK ok OK ok OK ok   OK ok   OK ok   OK ok) OK,
       PERFECT ITEM   10.000

(OK ok OK ok OK ok   OK ok   OK ok   OK ok) OK,
       PERFECT ITEM   10.000

(OK ok OK ok OK ok   OK ok   OK ok   OK ok) OK,
       PERFECT ITEM   10.000

(OK ok OK ok OK ok   OK ok   OK ok   OK ok) OK,
       PERFECT ITEM   10.000

(OK ok OK ok OK ok   OK ok   OK ok   OK ok) OK,
       PERFECT ITEM   10.000

(OK ok OK ok OK ok   OK ok   OK ok   OK ok) OK,
       PERFECT ITEM   10.000

(OK OK DUD dud OK ok   OK ok   OK ok   OK ok) DUD,
       FAULT SPOTTED, SECOND   8.000

(OK ok OK ok OK ok   OK ok   OK ok   OK ok) OK,
       PERFECT ITEM   10.000

(OK ok OK ok OK ok   OK ok   OK ok   OK ok) OK,
       PERFECT ITEM   10.000

(OK ok OK ok OK ok   OK ok   OK ok   OK ok) OK,
       PERFECT ITEM   10.000

OUTCOME OF SIMULATIONS

MEAN RECEIVED PER ITEM   9.600, SE   0.828
NUMBER OF SIMULATIONS   15.000
NUMBER OF PERFECTS   12.000   80.000%
NUMBER OF SECONDS   3.000   20.000%
NUMBER OF RTMS   0.000   0.000%

If you want more simulations with the same parameters, type GOTO 370 (return)
If you want to vary accuracy of panel inspection, type GOTO 270

---

5   An optimisation approach to the problem

We consider the problem of *maximising* $E(Y_1)$, the expected amount received per item, with respect to $\alpha_1, \ldots, \alpha_n$, the panel inspection rates, subject to $0 \leqslant \alpha_i \leqslant 1$ for each i. (In practice it may be impossible to attain $\alpha_i = 1$, so that the constraint $0 \leqslant \alpha_i \leqslant \max(\alpha_i)$ may be more realistic.)

Put $P^1 = P - S\alpha_F + F\beta_F$; this is the coefficient of δ_1 in (4.1). Then we seek to maximise, say,

$$f(\alpha_1, \ldots, \alpha_n) = P^1\delta_1 - C(\alpha_1, \ldots, \alpha_n) - \sum_1^n \rho_i \alpha_i \pi_i.$$

Now if n is large (say $n \geqslant 10$) and π_i small (say $\pi_i < 0.01$), then $\delta_1 \approx 1 - \sum_1^n \pi_i(1 - \alpha_i)$ is a reasonable approximation.

Of course it is impossible to make further progress in maximising $f(\alpha_1, \ldots, \alpha_n)$ without some assumptions about $C(\alpha_1, \ldots, \alpha_n)$, the cost of the panel inspection. Suppose therefore that

$$C(\alpha_1, \ldots, \alpha_n) = \sum_1^n C_i(\alpha_i),$$

where $C_i(\alpha_i)$ is the cost of inspecting panel i at accuracy α_i. Thus we assume that the cost of inspecting several panels is simply *additive* in the individual inspection costs. Then

$$f(\alpha_1, \ldots, \alpha_n) \approx g(\alpha_1, \ldots, \alpha_n) \quad \text{say,}$$

where

$$g(\alpha_1, \ldots, \alpha_n) = P^1 \sum_1^n \pi_i \alpha_i - \sum_1^n \rho_i \alpha_i \pi_i - \sum_1^n C_i(\alpha_i)$$

and so

$$\frac{\partial g}{\partial \alpha_i} = P^1 \pi_i \quad -\rho_i \pi_i \quad -\frac{dC_i}{d\alpha_i}$$

for $i = 1, \ldots, n$.

If we make a further simplifying assumption and take

$$C_i(\alpha_i) = k_i \alpha_i,$$

i.e. the cost of inspecting panel i depends *linearly* on the accuracy, then we see that $g(\alpha_1, \ldots, \alpha_n)$ is a *linear* function of $(\alpha_1, \ldots, \alpha_n)$, and

$$\frac{\partial g}{\partial \alpha_i} = (P^1 - \rho_i)\pi_i - k_i.$$

Thus, if $(P^1 - \rho_i)\pi_i - k_i > 0$, it is optimal to take α_i as near to 1 as possible while, if this quantity is negative, then we should take $\alpha_i = 0$, i.e.

do *not* inspect panel i before assembly if $P^1\pi_i < \rho_i \pi_i + k_i$.

Now the assumption that C_i is *linear* in α_i is probably unrealistic, since presumably, while $C_i(\alpha_i)$ is an increasing function of α_i, we would expect that

$$\frac{\mathrm{d}C_i}{\mathrm{d}\alpha_i} \text{ is } larger \text{ for } \alpha_i \text{ near } 0 \text{ or } 1 \text{ than for } \alpha_i \text{ near } 0.5.$$

However, the optimal solution given above may continue to be a good enough approximation even when we relax the condition of linear cost of panel inspection.

6   Suggestions for further work

(i) For a particular problem in which $C(\alpha_1, \ldots, \alpha_n)$ could be fairly well specified, it would be interesting to explore more rigorously the optimisation problem posed above.

(ii) In the approach outlined here, we assumed that all components were equally 'important' in determining whether the whole item was 'DUD' or 'OK'. In fact some components may be more critical than others; for example in a complicated machine we might find that the machine failed to work at all if certain key components were faulty, or perhaps worked not quite all of the time if certain other components were faulty. So far my approach has been to give all the components the same 'status' but, having established the basic notation, it would not be difficult to allow the components different relative statuses.

Similarly, my approach has been quite simple-mindedly *binary*; components, and items, are either 'DUD' or 'OK'. It would be interesting to generalise the problem to allow for varying degrees of faultiness in a component or an item.

(iii) In a practical problem, it may be fairly straightforward to assess *costs* such as P, S, F and ρ_i, but assessing fault rates π_i, particularly when they are known to be very small, may be a harder problem for the statistician. If the manufacturer is primarily interested in the *optimal* choice of $\alpha_1, \ldots, \alpha_n$ then it will not be necessary to estimate $\pi_1, \ldots, \pi_n$ with great precision. For example, we would probably take $\alpha_i = 0$ as long as we could be reasonably certain that

$$P^1\pi_i < \rho_i\pi_i + k_i,$$

i.e. $\pi_i < k_i/(P^1 - \rho_i)$.

However, in general the desired accuracy of estimation of the parameters (π_i) would probably depend on the desired accuracy in estimating $E(Y_1)$ and δ_1; this may not be very easy to handle.

Acknowledgement
I am very grateful to Marks and Spencer plc for the opportunity to work on this problem and to publish the results, and I would particularly like to thank Morrice Adelman and David Shaw for their time and trouble.

12

Statistical modelling of the EEC Labour Force Survey: a project history

M. AITKIN AND R. HEALEY

1   Background

From June 1981 to July 1984 the Centre for Applied Statistics (CAS) at the University of Lancaster held three one-year contracts with the Statistical Office of the European Communities in Luxembourg. The aim of these contracts was to determine whether mathematical models could be used to summarise the information on, and reduce the volume of tabulation of, unemployment rates from the biennial Labour Force Survey published by the Statistical Office.

The results of these studies are presented elsewhere (Aitkin and Healey, 1984b, 1985), and are fully documented in the contract reports on the three individual studies (Aitkin and Healey, 1982, 1983, 1984a). In this chapter we briefly describe the two contracting organisations, the terms of reference of the studies, how the studies were managed, and the use made of the results.

2   The Statistical Office of the European Communities

The Statistical Office of the European Communities (abbreviated hereafter to the multilingual acronym EUROSTAT) is a large directorate of the European Economic Community responsible for collating and disseminating information about the activities of the member states of the EEC. EUROSTAT appoints its staff from the Statistical Offices of individual member countries, and by direct recruitment for specialised positions.

EUROSTAT has the responsibility for collating the information collected by the central statistical offices of the EEC member countries, particularly that from the very large Labour Force, Structure of Earnings and Agriculture surveys; for ensuring the consistency of the definitions used as the basis of the national surveys; and for publishing the collated results through the Office of Official Publications, EUROSTAT's own publishing house. Since the readership of EUROSTAT's publications is drawn from all the member states, most of these publications have to be translated into at least the four main languages of the community: French, German, Italian and English.

Historically EUROSTAT has restricted itself to publishing 'factual' cross-tabulations of the results of large-scale surveys, and has left the task of interpreting the tabulated data to the reader.

3   The Centre for Applied Statistics

The Centre for Applied Statistics is a research and consulting centre of the University of Lancaster. Since its establishment in 1979 the Centre has been involved in a wide range of projects: the development of direct likelihood inference, applications of generalised linear models in medical and social statistics, the development and application of variance component models for clustered survey designs and school effectiveness studies, and the development of software for a wide range of applications.

The Centre provides statistical consulting services both inside and outside the University. A small number of staff are supported directly by the University; other staff are supported by Economic and Social Research Council (ESRC) research programmes and projects, by internal funding of specific data-analysis projects associated with survey work carried out by other departments, and by external contracts with organisations such as EUROSTAT.

4   Aims of the projects

The aims of the projects were to assess if and how the use of linear models could help with the interpretation of raw survey data, to investigate the feasibility of data reduction by using linear models, and to demonstrate the practicability of using such methods on large-scale survey data.

5   Reports and papers

The basic requirement for project reporting within each contract was to produce a draft report by the mid-point of each project year, and a final report by the end of each contract. The draft report for each year largely consisted of details of results from the first half of the contract period and a statement of intent for the work outstanding in the remainder of the contract. The final report reviewed the research carried out against the objectives set out in the work programmes and made recommendations for future work.

In addition to the formal reports made during the studies, a number of papers were produced both internally and externally and a number of seminars were given on both the statistical methodology and computational aspects of the work.

Work began in summer 1981 analysing French and Italian data from the 1979 Labour Force Survey, in the form of large contingency tables of counts of employed and unemployed individuals. These data had been extensively analysed by EUROSTAT and many of the interesting features in the data had already been identified, but were not made known to the Centre.

In November 1981 a seminar was held at EUROSTAT at which the overall methodology was reviewed and the first results presented. The final report on this project (H/81/46) showed that the modelling approach could reveal important features of the data, and patterns of age- and industry-related unemployment were identified. These patterns were very largely consistent with EUROSTAT's prior knowledge, except in a few cases which were attributed to data quality.

The second work programme for 1982–3 took data from the 1981 Labour Force Survey for France and Italy and demonstrated the stability of the French estimates over the two successive surveys. In addition, considerable effort was expended searching for methods of model fitting which could reduce computer processing time.

The third contract 1983–4 took data from the seven other member states of the community and attempted to identify a common model for the underlying structure of employment for each country, and to identify departures from the model in terms of the sampling designs used. This work was largely documented and discussed at a seminar held in Luxembourg in November 1983, together with the results of various other research projects aimed at analysing large-scale survey data. The proceedings of the seminar have subsequently been published (see Aitkin and Healey, 1984b) and a more technically comprehensive overview is given, in Aitkin and Healey (1985).

During the summer of 1983, work was completed on developing a version of the iterative scaling algorithm, required for fitting large models, for use on the ICL Distributed Array Processor (DAP). A copy of this program, developed in conjunction with the Centre's ESRC research programme, is now available through the DAP Support Unit at Queen Mary College, Mile End Road, London. The work leading to the development of the program was presented at the Parallel Computing 83 conference and is documented in Healey and Davies (1984).

The 1983–4 contract also involved the examination of methods of interpreting the parameter estimates derived from the models used, the assessment of goodness of fit through residual screening, and the development of simple graphical presentations for the results of the modelling.

6   Work programmes

Each contract had associated with it a work programme indicating the broad areas into which research should be directed. In general, these programmes did not require that specific analyses be performed, but suggested a possible class of analyses which could be investigated. None the less, the programmes formed a basis for the contracted work for the duration of each contract. The 1982–3 programme is reproduced below to show the level of direction contractually required by EUROSTAT. The day-to-day work of the project was subject to variation from these programmes, and we frequently responded to developments arising from EUROSTAT's own analyses of the data.

Work programme for Project H(82)70(228)

1. *Background*
The work conducted by the University of Lancaster for the Statistical Office of the European Communities has demonstrated on data from two member countries, France and Italy, the feasibility of fitting a model to partially aggregated data of one large and typical survey, namely the 1979 Labour Force Survey.

The work has shown that the results are interpretable and that the results are stable in that changes in the specification of some part of the model do not produce large changes in the estimates provided for other parts of the model. There exists therefore the possibility of making radical changes in the way in which data from this survey and probably other surveys are published. There are however a number of problems to be solved before this method could pass into routine use.

2. *The problems*
Before complete confidence can be given to the principle of reporting the parameters of a model rather than the cells of a table, it is essential that:

- the stability within any one survey should be demonstrated even more positively on the data of all member countries
- stability or near stability should be demonstrated over short to medium periods of time from one survey (1979) to the next (1981). It is known that structures do not normally change radically over short periods and it is important that such changes as are found should be interpretable in the light of knowledge of external changes. The 1981 data will become available after mid-summer 1982.
- the economics of computation and the choice of algorithms require further clarification. It is already known that the quantity of

computation required will exceed the effective capacity of the relatively small computer available to the University of Lancaster. Before any such procedures could pass into routine use, it would be necessary to integrate a version of the GLIM package, modified in the light of conclusions on the algorithms, into other software used on the Commission computers. This requires in the short run, the development of various GLIM macros, and subroutines. Longer run, it would mean interfacing with the SIGISE system.

3. The Centre for Applied Statistics of the University of Lancaster is therefore commissioned to extend its work on the 1979 and 1981 surveys. The objectives of this work are:

• to compare the models obtained on the 1979 and 1981 surveys for France and Italy, and to liaise with EUROSTAT and with the Centre d'Analyse Statistique des Structures et des Flux, Université de Paris X, in investigating the reasons for any instability in the models.

• to extend the analysis of the rate of unemployment, commenced on the 1979 data by deriving a uniform family of models for all Member States for the 1981 survey. These models will include the variables, age, sex, region, previous economic activity. They will, if it is practical, take into account the duration of unemployment. It is recognised that this last variable presents problems in the construction of models which it may be desirable to avoid by replacing unemploying rate as the dependent variable by the variable 'proportion of potential working time lost through unemployment'. The choice of dependent variable will be agreed between the Centre for Applied Statistics and EUROSTAT.

• (in so far as may be practical) to extend the analysis to such other aspects of the Labour Force data as may be agreed between the Centre for Applied Statistics and EUROSTAT. If it should not be feasible to incorporate data on the duration of unemployment into the model described in the previous paragraph, then a separate analysis of the structure of unemployment is likely to be the next priority.

• to advise EUROSTAT on the problems which may be met with the other aspects of the Labour Force Data and other surveys and specifically to give advice, if requested, relevant to the analysis of data of the Structure of Earnings Survey.

4. Because the quantity of computation likely to be required is beyond the resources of the computer available to the Centre for Applied Statistics, and because the long-term intention is that facilities for running this work are required on the computers of the Commission, the University of Lancaster is requested to carry out the bulk of this work using terminals to access the 2900 computers of the

Commission at Luxembourg via EURONET and to develop the procedures by which this work could be handled later by staff of EUROSTAT.

7   Guidelines for consultancy projects

As with any large organisation, EUROSTAT has set standards by which external consultancy projects are run. These standards are provided to contractors as guidelines. The guidelines provide a framework within which contractors are expected to carry out their work. They cover personnel, meetings with the Commission, reports, permits and licenses, secrecy, liability, results and royalties, subcontracting, non-performance, resolution of disputes and amendments to the contract.

8   Contract details

Contract payments were tied to the delivery of the interim and final study reports. EUROSTAT reserved the right to ask for further clarification of points raised in the reports within one month of receipt of the final report. The structure of the payments was 30% at the start of the contract, 30% on receipt of the draft report and 40% on the acceptance of the final report.

Payments for the first two contracts were made in pounds sterling, and for the final contract in the Commission's internal accounting unit (ECU). Fluctuations in currency rates made some change in the overall value of the last contract in local currency.

In addition to the fixed price of the contract, the Centre was reimbursed for the cost of computer time and for travel and subsistence when such travel was made at the Office's request.

Over the three years of the contracts their value to the University was £83 000. The contracts paid for one full-time research fellow, the part-time secondment of one permanent Centre staff member and Centre and University overhead costs. Temporary consulting staff were appointed to cover the secondment of the permanent staff member.

9   Hardware available

The three projects led to the development of much software, not least of which was a communications package written by the Centre Assistant Director, Brian Francis, to assist in the transfer of data between various machines used.

In the first year of the project the University's ICL 2960 computer was used running under the DME 2900 operating system. Subsequently the Commission's ICL 2982 running under VME/B was used, and the

ICL DAP at Queen Mary College and an ACT SIRIUS 1 at Lancaster which allowed dial-up access to all of the other machines. In the later stages of the final contract some theoretical aspects of the study were investigated using the University's VAX 11/780 running under VMS.

10   Monitoring performance

One of the more difficult aspects of working in a research environment is the problem of setting realistic project targets. All three of the contracts held were 'open-ended' in the sense that there were always aspects of the work programmes where more effort could be expended.

The short-term nature of the contracts required a firm determination to concentrate on key aspects of the programmes and to use the other elements of the programmes as back-up tasks when a major obstacle was encountered or as supplementary work towards the end of the contract.

The strategy adopted was to pursue a number of main strands of research and to switch between strands in the event of unforeseen difficulties such as machine unavailability, or when clarification of particular points was required from someone who was unavailable.

As in many projects with a large research component, the amount of effort to be put into each task could not be easily estimated in advance, and following on from this the attendant milestones could only be guessed at.

This created some problems in project control which were overcome by setting out each chapter in the final report as a separate task and keeping a 'rolling' draft report constantly updated as work on each task progressed. At the end of each contract only a minimum of effort was required to edit the rolling draft report and to add final conclusions and a management summary.

11   Assessing completion

As with many research projects, assessing completion of the project is a rather arbitrary process. The many project deadlines required us to draw the line for the current work as the completion date for each final report arrived. It is in the nature of this kind of project that some avenues are only partially explored, if they are not omitted entirely.

12   Personnel

During the three years over which the contracts ran, a roughly similar staffing structure was maintained. The structure was a director responsible for deciding the priorities to be assigned to the various strands of the work programme and also acting as a technical advisor on the more difficult theoretical problems; working for and responsible to the director was a research fellow who was responsible for implementing policy decisions agreed with the director and for the day-to-day running of the project and documentation of the analyses carried out. In addition to this permanent establishment, other members of the Centre were available on an *ad hoc* basis both formally and informally.

13   Working on short-term contracts

On all three contracts the appointment of the research fellow was based on a fixed one-year contract. As the subsequent contracts depended on the acceptance and implementation of the previous year's results, there was considerable pressure during the period immediately before the end of a contract to seek fresh employment on a 'safety net' basis.

This form of employment was injurious not only to the research fellow but also to the Centre, as in the longer term better prospects must arise outside the Centre. Recently, the Centre has been able to adopt a more flexible policy of employing a research fellow on the expectation of obtaining funding externally to the University. This at least provides some security but less than could be offered by, say, a rolling two-year contract.

14   Variety of research contracts

One of the major advantages to a department structured like the Centre is the opportunity for cross-fertilisation of ideas by researchers working in different but related areas. Thus the more pressing technical problems can be talked through, and a wide range of opinions and analyses can be brought to bear on how best to attack these problems.

15   Conclusions

(i) In as far as the aims of the projects were to show that modelling of large-scale survey data is feasible, and that the results of the fitted models may be easily interpreted, the project's aims were successfully realised.

The longer-term use of the conclusions of these studies can only be guessed at, but some indication is given by the proceedings of the EUROSTAT seminar on analysing large-scale data sets. EUROSTAT plays a complex and political role in its delicate job of trying to influence the various Statistical Offices of the Community to provide its raw data in a form that has a common base. Further, EUROSTAT is limited in its resources, and a major commitment to enhancing its output at the cost of additional computation will take some time to evaluate.

At least the studies showed those who were tabulating raw data at EUROSTAT that modelling can produce results which help to highlight anomalies in the data and may provide simple but more informative descriptions of the data, which may be valuable for many users wanting more detailed tabulations than currently produced.

(ii) These projects provided the Centre with an interesting research area whose results could have great practical benefit. In addition, the projects provided funds to strengthen the establishment of the Centre during a critical period of growth.

References
Aitkin, M. and Healey, R. (1982) Report to EUROSTAT on study H/81/46.
Aitkin, M. and Healey, R. (1983) Report to EUROSTAT on study H82(78).
Aitkin, M. and Healey, R. (1984*a*) Report to EUROSTAT on study H83(87).
Aitkin, M. and Healey, R. (1984*b*) Mathematical modelling of the EEC Labour Force Survey. In *Recent Developments in the Analysis of Large-Scale Data Sets*, Luxembourg: Office for Official Publications of the European Communities.
Aitkin, M. and Healey, R. (1985) Statistical modelling of unemployment rates from the EEC labour force survey. *J.R.Statist.Soc.* A. **148**, 45–56.
Healey, R. and Davies, S. (1984) Statistical model fitting on the ICL Distributed Array Processor. In *Parallel Computing 83*, North-Holland.

Bibliography on statistical consulting

D. J. HAND

(Preparation of this bibliography was considerably eased by the earlier work of Woodward and Schucany, 1977.)

Altman, D. (1982) Statistics and ethics in medical research: VIII. Improving the quality of statistics in medical journals. *British Medical Journal*, **282**, 44–7.

Bancroft, T. A. (1971) On establishing a university-wide statistical consulting and cooperative research service. *The American Statistician*, **25**, 21–4.

Bancroft, T. A. (1972) On teaching of service courses in statistics. *The American Statistician*, **26**, 14–16.

Bancroft, T. A. and Kennedy, W. J. (1972) Consultation and education programs in statistical computing for colleges and universities. *Proceedings of the Computer Science and Statistics Sixth Annual Symposium on the Interface*, American Statistical Association, pp. 24–7.

Barnett, V. D. (1976) The statistician: Jack of all trades, master of one? *The Statistician*, **25**, 261–79.

Baskerville, J. C. (1981) A systematic study of the consulting literature as an integral part of applied training in statistics. *The American Statistician*, **35**, 121–3.

Bliss, C. I. (1969) Communication between biologists and statisticians, a case study. *The American Statistician*, **23**, 15–20.

Boardman, T. J. (1969) Letter to the editor. *Biometrics*, **25**, 434.

Boen, J. R. (1972) The teaching of personal interaction in statistical consulting. *The American Statistician*, **26**, 30–1.

Boen, J. R. (1982) A self-supporting university statistical consulting center. *The American Statistician*, **36**, 321–5.

Boen, J. R. and Smith, H. (1975) Should statisticians be certified? *The American Statistician*, **29**, 113–14.

Boen, J. R. and Zahn, D. A. (1982) *The Human Side of Statistical Consulting*, Belmont, California: Wadsworth.

Bross, I. D. J. (1974) The role of the statistician: scientist or shoe clerk. *The American Statistician*, **28**, 126–7.

Cameron, J. M. (1969) The statistical consultant in a scientific laboratory. *Technometrics*, **11**, 247–54.

Committee on Training of Statisticians for Industry (1980) Preparing statisticians for careers in industry (with discussion). *The American Statistician*, **34**, 65–75.

Court, A. T. (1952) Standards of statistical conduct in business and in government. *The American Statistician*, **6**, 6–14.

Cox, C. P. (1968) Some observations on the teaching of statistical consultancy. *Biometrics*, **24**, 789–802.

Daniel, C. (1969) Some general remarks on consultancy in statistics. *Technometrics*, **11**, 241–6.

Daniels, H. E. (1975) Statistics in universities – a personal view (with discussion). *Journal of the Royal Statistical Society*, Series A, **138**, 1–17.

Deming, W. E. (1965) Principles of professional statistical practice. *Annals of Mathematical Statistics*, **36**, 1883–1900.

Deming, W. E. (1966) Code of professional conduct. *Sankhya* B, **28**, 11–18.

Deming, W. E. (1972) Code of professional conduct: a personal view. *International Statistical Review*, **40**, 215–19.

Eisenhart, C. (1947) The role of a statistical consultant in a research organisation. *Proceedings of the Third International Statistics Conference*, pp. 309–13.

Feinstein, A. R. (1970) Clinical biostatistics VI: Statistical malpractice – and the responsibility of the consultant. *Clinical Pharmacology and Therapeutics*, **11**, 898–914.

Feller, W. (1969) Are life scientists overawed by statistics? *Scientific Research*, **4**, 24–7.

Finney, D. J. (1968) Teaching biometry in the university. *Biometrics*, **24**, 1–12.

Finney, D. J. (1982) The questioning statistician. *Statistics in Medicine*, **1**, 5–14.

Freeman, W. W. K. (1963) Training of statisticians in diplomacy to maintain their integrity. *The American Statistician*, **17**, 16–20.

Gale, W. A. (1985) (ed.) *Artificial Intelligence and Statistics*, Reading, Massachusetts: Addison-Wesley.

Gibbons, J. D. (1973) A question of ethics. *The American Statistician*, **27**, 72–6.

Gore, S. M., Jones, I. G. and Rytter, E. C. (1977) Misuse of statistical methods: critical assessment of articles in BMJ from January to March 1976. *British Medical Journal*, **1**, 85–7.

Greenfield, A. A. (1979) Statisticians in industrial research: the role and training of the industrial consultant. *The Statistician*, **28**, 19–27.

Griffiths, J. D. and Evans, B. E. (1976) Practical training periods for statisticians. *The Statistician*, **25**, 125–8.

Hand, D. J. (1984) Statistical expert systems: design. *The Statistician*, **33**, 351–69.

Hand, D. J. (1985) Statistical expert systems: necessary attributes. *Journal of Applied Statistics*, **12**, 19–27.

Hand, D. J. (1985) The role of statistics in psychiatry. *Psychological Medicine*, **15**, 471–6.

Harshbarger, B. (1968) Teaching of statistical consulting. *Biometrics*, **24**, 455 (Abstract).

Healy, M. J. R. (1973) The varieties of statistician. *Journal of the Royal Statistical Society*, Series A, **136**, 71–4.

Hooke, R. (1980) Getting people to use statistics properly. *The American Statistician*, **34**, 39–42.

Hunter, W. G. (1981) The practice of statistics: the real world is an idea whose time has come. *The American Statistician*, **35**, 72–6.

Hyams, L. (1969) Letter to the editor. *Biometrics*, **25**, 431–4.

Hyams, L. (1971) The practical psychology of biostatistical consultation. *Biometrics*, **27**, 201–12.

Kanji, G. K. (1979) The role of projects in statistical education. *The Statistician*, **28**, 19–27.

Kastenbaum, M. A. (1969) The consulting statistician: who needs him? *Oak Ridge National Laboratory Review*, pp. 9–11.

Kimball, A. W. (1957) Errors of the third kind in statistical consulting. *Journal of the American Statistical Association*, **57**, 133–42.

Lurie, W. (1958) The impertinent questioner: the scientist's guide to the statistician's mind. *The American Scientist*, **46**, 57–61.

Marquardt, D. W. (1979) Statistical consulting in industry. *The American Statistician*, **33**, 102–7.

Marquardt, D. W. (1981) Criteria for the evaluation of statistical consulting in industry. *The American Statistician*, **35**, 216–19.

Mead, R. (1976) Statistical consulting in a university. *The Statistician*, **25**, 213–18.

Miller, R. G. Jr, Efron, B., Brown, B. W. Jr and Moses, L. E. (1980) *Biostatistics Casebook*, New York: Wiley.

Morton, J. E. (1952) Standards of statistical conduct in business and government. *The American Statistician*, **6**, 6–7.

Moses, L. and Louis, T. A. (1984) Statistical consulting in clinical research: the two way street. *Statistics in Medicine*, **3**, 1–5.

Mosteller, F. (1971) Report of the evaluation committee on the University of Chicago Department of Statistics. *The American Statistician*, **25**, 17–24.

Pregibon, D. and Gale, W. A. (1984) REX: an expert system for regression analysis. *COMPSTAT-84*, Prague.

Salsburg, D. S. (1973) Sufficiency and the waste of information. *The American Statistician*, **27**, 152–4.

Schucany, W. R. (1972) Some remarks on educating problem solvers. *Proceedings of the Computer Science and Statistics Sixth Annual Symposium on the Interface*, American Statistical Association, pp. 33–6.

Snyder, M. (1972) Letter to the editor. *The American Statistician*, **26**, 59.

Sprent, P. (1970) Some problems of statistical consultancy (with discussion). *Journal of the Royal Statistical Society*, Series A, **133**, 139–65.

Sterling, T. D. (1973) The statistician vis-a-vis issues of public health. *The American Statistician*, **27**, 212–17.

Tarter, M. and Berger, B. (1972) On the training and practice of computer science and statistical consultants. *Proceedings of the Computer Science and Statistics Sixth Annual Symposium on the Interface*, American Statistical Association, pp. 16–23.

Urquhart, N. S. (1972) On consultation and education near the interface. *Proceedings of the Computer Science and Statistics Sixth Annual Symposium on the Interface*, American Statistical Association, pp. 28–32.

Watts, D. G. (1970). A program for training statistical consultants. *Technometrics*, **4**, 737–40.

White, S. J. (1979) Statistical errors in papers in 'The British Journal of Psychiatry'. *British Journal of Psychiatry*, **135**, 336–42.

Woodward, W. A. and Schucany, W. R. (1977) Bibliography for statistical consulting. *Biometrics*, **33**, 564–5.

Yates, F. and Healy, M. J. R. (1964) How should we reform the teaching of statistics? *Journal of the Royal Statistical Society*, Series A, **127**, 199–210.

Zahn, D. A. and Isenberg, D. J. (1979) Non-statistical aspects of statistical consulting, *Proceedings of the American Statistical Association, Section on Statistical Education*, 67–72.

Zahn, D. A. and Isenberg, D. J. (1983) Nonstatistical aspects of statistical consulting. *The American Statistician*, **37**, 297–302.

Zelen, M. (1969) The education of Biometricians. *The American Statistician*, **23**, 14–15.

Name index

Adcock 35
Adelman 170
Aitkin 77, 171, 173
Alpert 62
Altham 138, 161, 162, 163
Altman 8, 65
Aranda-Ordaz 139
Armitage 7, 43

Baker 77
Barnett 26, 31, 34, 36, 37, 38, 39, 110, 111, 112, 113, 114, 115, 116, 117, 129, 130
Bartlett 31, 94
Bascombe 110
Bathen 45
Bennett 142
Berger 7
Birch 36
Boddy 148
Boen 7, 9
Box 51
Boyce 139
Briese 121, 123, 126, 127, 131, 132
Brooks 138
Brown 83, 114
Buhagiar 112

Carter 139, 142
Christensen 68
Clayton 77
Cloke 62
Cochran 51, 96
Conover 51
Cormack 109
Cox 4, 51, 61, 65, 96, 142

Dallwitz 111, 114
Daniel 5
David 77
Davies 173

Descartes 20
Diggle 139, 142, 143
Dighton 83, 107
Dixon 113, 114, 146
Dolby 29, 40
Dorff 33
Dunn 73, 74, 77

Elashoff 68
El-Sayyad 40
Epstein 60, 62
Evans 7
Everitt 73, 145

Farewell 143
Federer 45
Feinstein 9
Felsenstein 110
Finney 9
Fisher 6
Fogarty 73, 76
Francis 176
Freeman 134

Gale 9
Geary 31
Geier 121, 131, 132
Gilmour 110
Gladen 139
Gong 67
Good 16
Goodhardt 6
Gore 8
Gould 147, 148
Gower 110, 113, 114
Gratton 139, 142, 143
Griffiths 7
Gurland 33

Hall 111

Subject index